D1066055

DATA COMMUNICATIONS TECHNIQUES AND TECHNOLOGIES

DATA COMMUNICATIONS TECHNIQUES AND TECHNOLOGIES

Joel Effron

Lifetime Learning Publications • Belmont, California

A division of Wadsworth, Inc.

London, Singapore, Sydney, Tokyo, Toronto, Mexico City

Jacket and Text Designer: Joe di Chiarro

Illustrator: Inge Infante

Compositor: Graphic Typesetting Service

Printed in the United States of America

1 2 3 4 5 6 7 8 9 10—87 86 85 84

Library of Congress Cataloging in Publication Data

Effron, Joel
Data communications techniques and technologies.

Bibliography: p.
Includes index.
1. Data transmission systems. I. Title.
TK5105.E34 1984 384 84–852
ISBN 0-534-03270-2

CONTENTS

The terms defined relate to the equipment, to how the equipment is connected in order for data communications to take place, and to transmission codes and modes.

This chapter looks at the rules of data communications—called protocols—and describes some of the most common protocols for data link control and error detection.

This chapter describes different types of modems and the related "black boxes" that make data communications possible.

LIST OF ILLUSTRATIONS

Figure

PREFACE

PURPOSE OF THE BOOK

This book is intended to be an easy-to-understand, up-to-date introduction to *data communications* that covers all the key aspects of its subject. The book deals primarily with the following:

- Concepts
- Vocabulary
- Approaches to making informed hardware, software, and network decisions
- Discussion of numerous data communications problems and their solutions
- Evaluating data communications vendors
- Establishing trouble-shooting procedures.

AUDIENCE

This book is intended for professionals in management, operations, and technical areas who are increasingly relying on data communications to help improve their daily performance. This includes those in:

- Data processing, who will learn how data communications relates not only to mainframe applications but more importantly to the rapidly growing base of mini- and microcomputer systems
- Office automation, who will learn how data communications is an integral part of electronic mail and message systems, word processing, information storage and retrieval systems; and who will also see how private branch exchanges (PBXs), dedicated local-area networks (LANs), security systems, and energy management systems relate to office automation.
- Telecommunications, who will learn about the merging of voice and data communications, and how PBXs can be one of the answers to a company's data communications needs.

ORGANIZATION

This book is organized logically so as to lead you through the data communications maze one step at a time. It introduces basic concepts, definitions, and vocabulary right at the beginning, and then proceeds to examine more advanced concepts. After the basic concepts and definitions have been dealt with, the balance of the book is arranged in a modular fashion such that you can, if you wish, study only those chapters or groups of chapters that interest you most.

FEATURES

Specific features of this book include the following:

- Easy-to-understand and in-depth explanations of data communications concepts, devices, and vocabulary.
- Many useful illustrations specifically designed to clarify the more complicated aspects of data communications.
- A complete and in-depth glossary, each term of which is both cross-referenced in the text and also indexed.
- A detailed section on economic analysis showing how to make a selection of equipment or services with different payment terms.
- Plenty of practical suggestions on how to solve data communications problems and save money; for example, how to take a standard four-wire private line that most companies use for data communications and create two complete circuits on that line, thus doubling the amount of data you can send.
- Methods you can use to transmit information from one place to another and still be sure that the information you receive is the same as the information you transmitted.
- The reorganization of AT&T and how to adapt to it.
- Alternate approaches to data transmission, such as fiber optics, microwave, infra-red, cellular radio, and how to use the AC power wiring already in your building for data communications.

When you have finished reading this book, you will not only be able to read and understand the data communications literature, but be able to converse intelligently with vendors, select the data communications hardware, software, and services you need, implement money-saving techniques, and improve the quality of your entire data communications system.

ACKNOWLEDGMENTS

Much more than mere thanks go to my wife, JoAnn, who put up with my eccentricities, and to my daughter Amanda, who was deprived of many hours of time that I should have been spending with her. A very special word of thanks goes to Perry Edwards, a dear friend and author of numerous books on data processing who talked me into writing this book. A final thank you goes to the many friends who came up with various facts and figures for use in the book, who helped with ideas and suggestions, and, most of all, who offered constant encouragement.

Joel Effron

DATA COMMUNICATIONS TECHNIQUES AND TECHNOLOGIES

1

WHAT IS DATA COMMUNICATIONS?

On your way to work, you were thinking about George Orwell's *1984* (which you finished reading last night), musing over which of the predicted events had occurred and which ones had not. But now you are entering your office and *1984* seems far behind you as you see the flashing red message light on your office information console.

You sit down at your desk and call up the messages: your boss has scheduled two appointments for you this morning; there is a reminder about this morning's video conference; the client from New York wants you to call; and there are twenty nonpriority messages as well, plus a voice message from home.

You quickly play the voice message while instructing the computer to retrieve the file on the New York client. The message comes through: "Don't forget to pick up steak for dinner tonight."

The telephone rings; you answer it and carry on a hands-free conversation over the speaker phone while you review the client file on the computer screen. You instruct the computer to prepare some forecasts complete with graphics—pie charts and bar graphs are very effective, you find. You tell the computer to get the industry numbers from the data base in New York that you contracted with last week.

While you continue your phone conversation, the computer initiates another phone conversation with the New York data bank and gets the information you've requested.

The computer comes back with the forecasts. You read them while you continue with your conversation on the speaker phone. They look good, but you need hard copy of those graphs and charts so you push a button to print them on your department's laser printer.

Your chime sounds; you put your call on hold while your computer tells you that it is time for the video conference. You excuse yourself from your phone call, send the results of your computerized forecasts to your boss's computer terminal with a priority message that you should discuss them ASAP, and you adjust your clothes and hair just in time for the television camera in your office to focus on you for the start of the video conference.

The conference is over. It's peaceful in your office; all you can hear is the quiet country music playing in the background. Your computer selected it just for you, based on the taste profile you entered into the computer months ago, by accessing that new computer-selectable custom music service in Nashville with an ordinary dial-data phone call.

Your mind drifts back to *1984,* and you smile when you realize that you have very routinely handled simultaneous voice, data, and electronic messages, warning signals, and hard-copy picture transmission, as well as background music and a video conference, all from your desktop information console and all over the same wires that just a few short years ago handled only a telephone conversation. In those few years, your office has moved into the electronic communications age.

This is office automation and a glance at the electronic communications explosion. It's coming soon to your nearest office! Are you ready for it? Do you know, for example, whether the communications described in that story consisted of voice, data, or video communications? And were they transmitted in analog or digital form? This book provides the answers to these questions, giving you the information you need in order to take advantage of the latest developments in data communications.

In this chapter we lay the groundwork for understanding these developments by defining data communications, distinguishing it from voice communications, and by describing some of its applications.

A PRACTICAL DEFINITION OF DATA COMMUNICATIONS

The process of data communications involves taking data from a device such as a computer terminal, transmitting or sending that data to a receiver such as a printer or another terminal, and then storing or displaying the data. For example, if a user is sending a letter from his or her word processor to a word processor in another city, the data communications process allows the data (in this case, the letter) to be taken from the disk on which the letter is stored, transmitted to the remote word processor, and then stored on the disk of the receiving word processor from which the letter can be printed. Or, if a user is entering information into the keyboard of a computer terminal, the data communications process "takes" that information from the keyboard and transmits it to the main memory of the computer; from there the information will be processed, stored on disk, or displayed.

If two computers are sending information to each other, or if a computer terminal is receiving information from a host computer, or if a device on a spaceship is sending information on the temperature in space back to Earth, we call that data communications.

Keeping the above examples in mind, we can now formulate a simple and practical definition of data communications: *data communications* is the process of transmitting information that originates with or is generated by a machine, typically a computer or a computer terminal, and is intended to be received by another machine.

DATA COMMUNICATIONS VERSUS VOICE COMMUNICATIONS

Data communications is often talked about in connection with voice communications, since voice communications media (such as telephone lines) are often used for transmitting data from one machine to another. Let's see how these terms relate, and clear up the confusion you are likely to encounter in data communications literature.

Defining Voice Communications

The process of voice communications, especially when conducted over the telephone, is familiar to all of us. A person talks into the telephone transmitter, which sends the voice information to the telephone receiver at the other end where the voice can be listened to. Although voice communications usually occurs between two humans, we also call it voice communications when a computer uses a synthesized "voice" to talk to a human.

In many ways, voice and data communications are analogous; that is, machines use data communications to talk to each other much in the same way that humans use voice communications (for example, the telephone system) to talk to each other.

We can now formulate the following simple definition of voice communications: *Voice communications* is the process of transmitting information that originates as a human voice, or as an imitation of the human voice, and that is intended to be received by another human or by computerized voice-recognition equipment.

The Confusion Surrounding the Term "Data Communications"

We have now presented two of the most important definitions we will work with in this book, but you should know that there is a certain amount of arbitrariness in these definitions and that they may conflict with definitions used by other authors. For example, some authors insist that data communications must be digital and that analog data transmission between two machines does not qualify as "data communications" ("digital" and "analog" are defined briefly below and discussed in depth in the next chapter).

Other authors say that if a voice conversation between two people is digitized before being transmitted, then that voice conversation becomes data communications. I still call that conversation "voice communications," based on the definitions we will use in this book.

Other authors insist that data communications occurs only if a computer is involved. Some exclude facsimile or FAX (the transmission of pictures electronically) from data communications. In other words, there is no universal agreement on the meaning of this term.

Nevertheless, I believe you will find my definitions of voice and data communications very useful and clear enough to avoid any confusion. This book will also show you how voice and data communications will tend to blend together in the future, perhaps eliminating these worries about different definitions.

Digital and Analog Transmission

The confusion surrounding the use of the terms voice and data communications can be blamed largely on what is called "digital transmission."

Digital transmission is a technique whereby information (whether it is data, voice, or video) is converted to, or coded as, a series of 0's and 1's before being transmitted. These 0's and 1's are called "binary digits," hence the term *digital* transmission. By using digital transmission, voice, data, and video information can all be combined together and transmitted over the same wires.

Analog transmission, on the other hand, involves the modulation of an information carrier (such as an electronic pulse, a constant voltage, or a radio signal) so that it represents, or becomes the *analog* of, the information being sent (such as a sound wave). The height (or amplitude) of a pulse, for example, might be modulated so that it matches the variations in the height of the sound wave; the pulse, then, acts as the analog of the sound wave.

Analog transmission is often used for voice communications, but the two terms are not synonymous. Similarly, binary digits or 0's and 1's are the natural language of computers and computer terminals. Remember that, when computers communicate with each other or with other devices like printers or computer terminals, we call the process data communications. Since the data are normally stored in the computer as digital information, some people use the term data communications to mean any form of communications that involves digital transmission.

Chapter 2 discusses the relationship of the *mode* of transmission (digital or analog) to the *signal* being transmitted (voice or data) and describes the blending of voice and data communications in greater depth. But what you need to know now is that, although this book concentrates on data communications per se, much of what is discussed also applies to the digital transmission of voice and video.

DATA COMMUNICATIONS APPLICATIONS

Now that we have a simple working definition of data communications and have distinguished it from digital transmission, let us look at various applications for data communications. First, we will look back briefly at applications from the early history of data communications; then we will examine the applications that are common today; and finally we will look at what we expect the future holds for data communications.

As we describe various applications, look for the ones that your work involves. This will help you focus on the sections of this book that are most practical for you.

Early Data Communications Applications

Telegraphy The earliest significant example of data communications was the telegraph. Even to this day some of the jargon of telegraphy can be found in the modern data communications vocabulary. For example, "mark" and "space" (which we will define in Chapter 3) are terms that originated with telegraphy and are still in use today in conjunction with computer-to-computer communications, although their meaning has changed somewhat. The telegraph, telex, and TWX (telex and TWX are dial-up teletype networks offered by Western Union) are all examples of data communications that have been with us for decades (see Figure 1-1).

Communication Between Computers and Peripheral Devices The connection of peripheral devices (such as computer terminals, printers, disk drives, or tape drives) to their host computers is the most common example of data communications involving computers (see Figure 1-2).

Perhaps the most significant early development in data communications occurred when people wanted to take a computer terminal that was designed to operate within fifty feet of the computer and locate that terminal across town, several miles from the host computer, say at a branch office (see Figure 1-3). Developing ways to transmit data over distances of more than a few hundred feet paved the way for the long-distance applications that are among the most important aspects of data communications today.

Common Data Communications Applications Today

Facsimile Transmission An example of a current data communications application that has grown rapidly is facsimile transmission, also known as FAX. In this application, pictures of any document are transmitted over telephone lines to a receiving facsimile machine, which can be located in the same building,

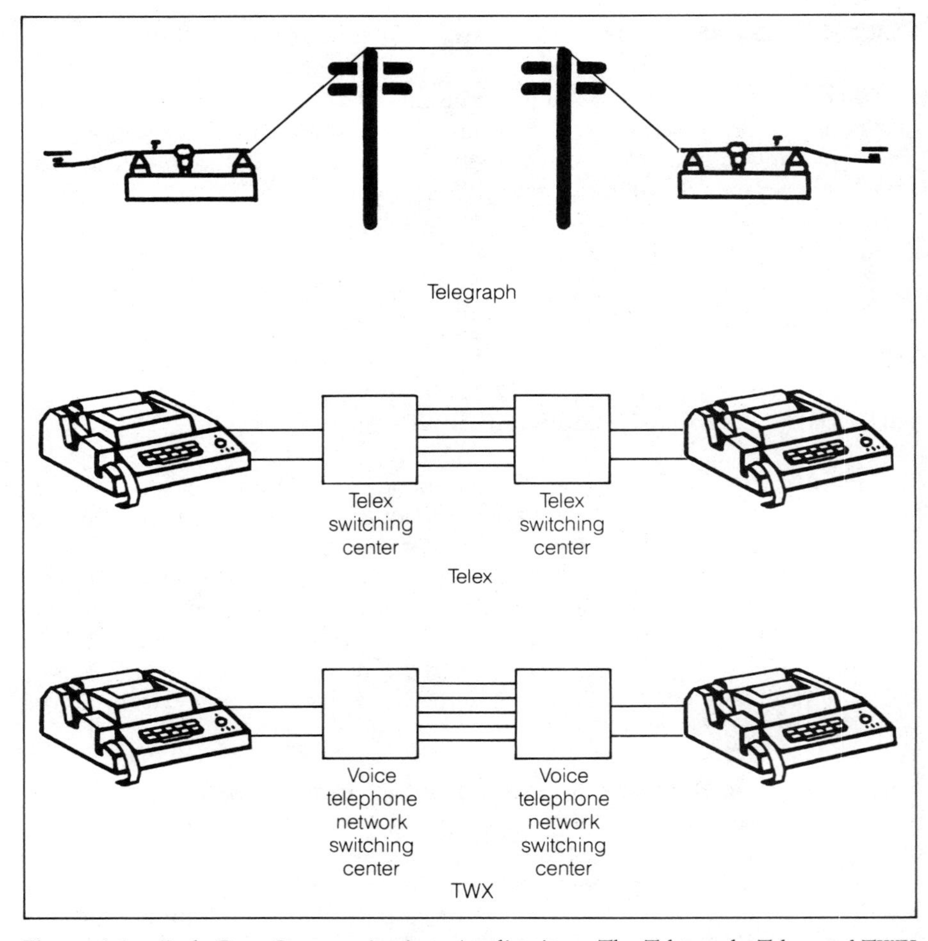

Figure 1-1. Early Data Communications Applications: The Telegraph, Telex and TWX

miles away, or even in another country (see Figure 1-4). As we mentioned earlier, some authorities separate FAX from data communications, but this book treats FAX as a type of data communications.

Distributed Processing This is a system whereby several computers, usually mini- and microcomputers, are connected in a network. In some networks, each computer can "talk" to other computers in the network and thus have access to the data base and peripherals (such as the printers or tape drives) of the other computers; in others, the remote computers access the data base and peripherals of a central-site mainframe computer (see Figure 1-5). Distributed processing is

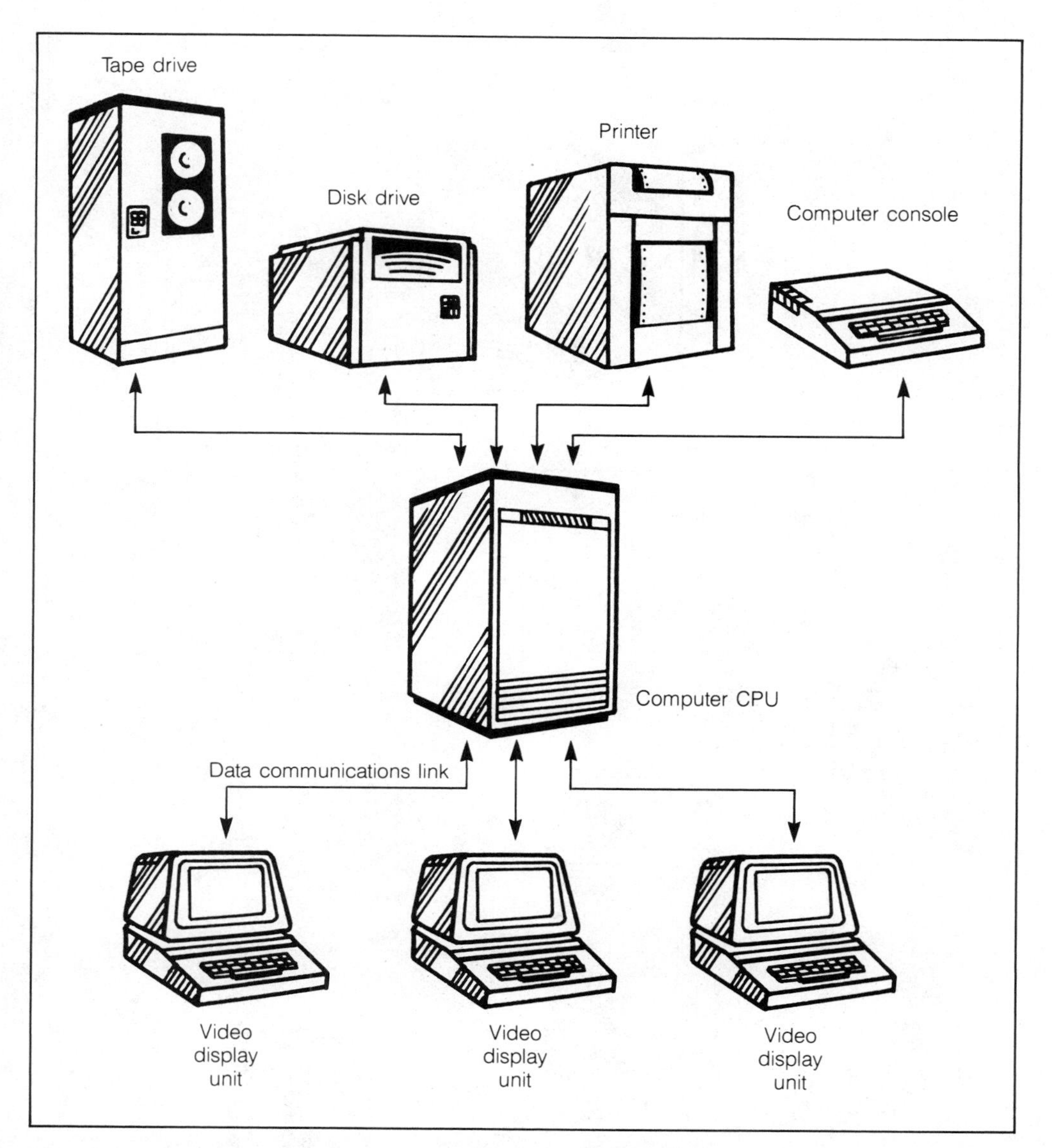

Figure 1.2. How Data Communications Links a Computer to Its Peripheral Devices

becoming very popular as inexpensive computers and reliable data communications make it a practical choice for many companies.

A common implementation of distributed processing is to have a central data base (such as an inventory file for a parts distribution company) on disk

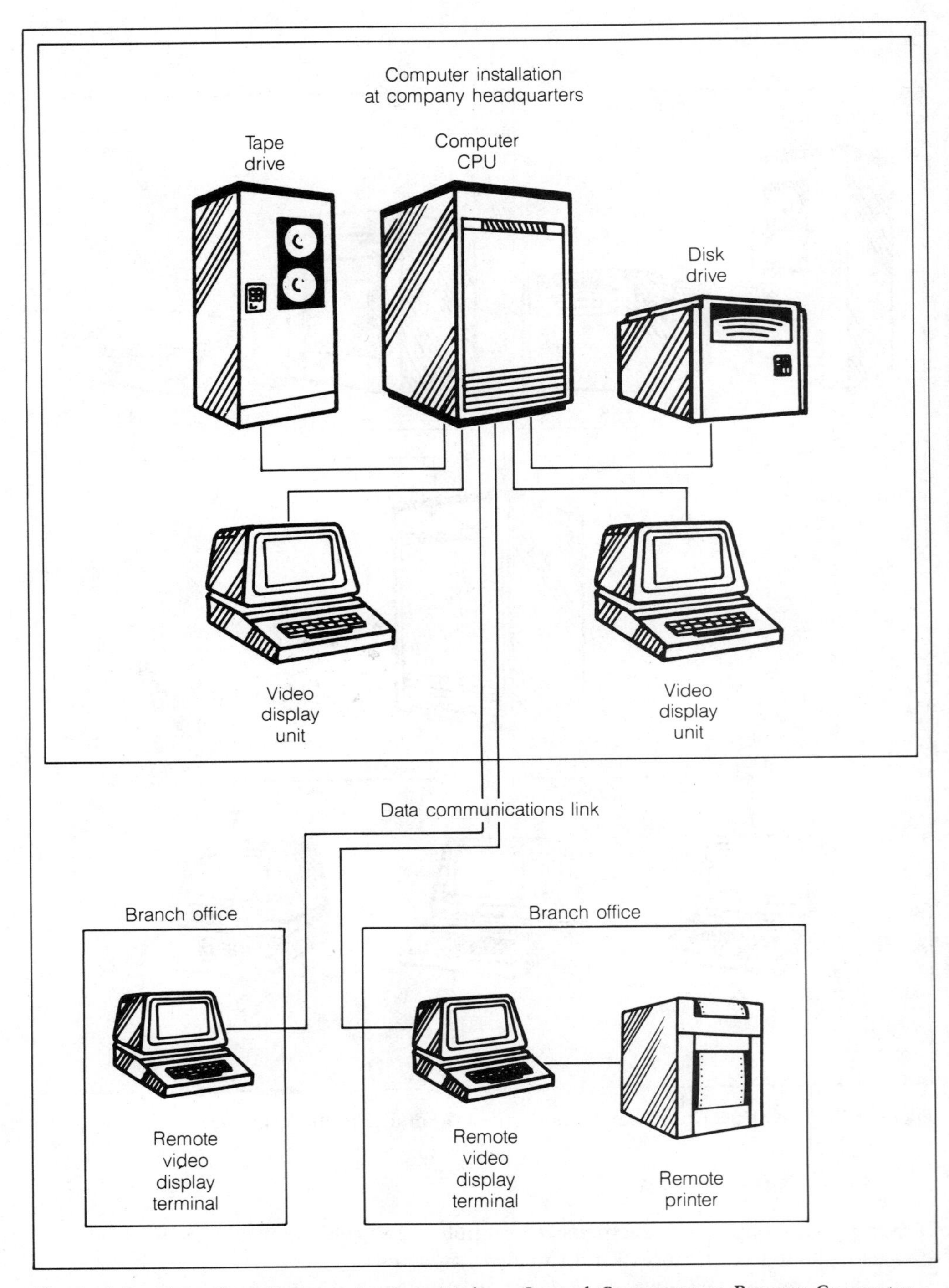

Figure 1-3. How Data Communications Links a Central Computer to Remote Computer Terminals

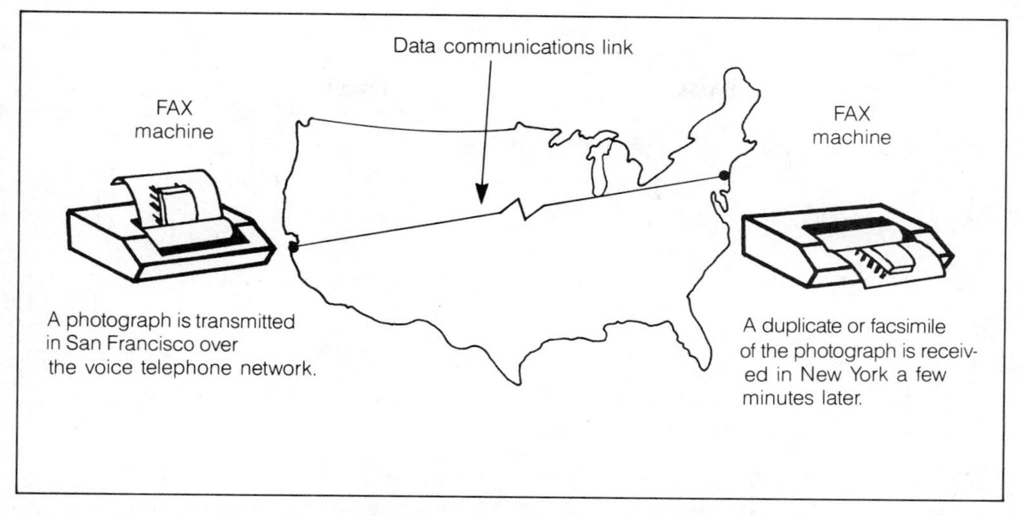

Figure 1.4. **Data Communications and Facsimile Transmission**

storage that can be accessed by many small computers located throughout the company. Each computer can access and update records in the central data base (for example, to check for availability or to add shipping and receiving information). These same computers can also have long reports (such as inventory valuation reports) printed automatically at the central site. This eliminates the need for high-speed printers and massive disk storage at each of the remote computer locations, while still allowing each computer access to these expensive resources.

Another simple example of this process is where a number of word processors, each having its own local storage and low-speed printer facilities, also all have access to a central data base (for retrieval or storage of letters) or to a central high-speed laser printer.

Distributed processing by its very nature requires significant data communications capability to make it work.

Networks of Remote Terminals Another type of network that requires significant data communications capability consists of remote terminals connected to a central computer. (In distributed processing, remember, remote *computers,* which have more processing power than most terminals, are connected to each other or to a central mainframe computer.) Banks, retail stores, and the airlines are all large users of this type of data processing.

The banking industry is adding automatic teller machines throughout the world and connecting them with the bank's central computers. A customer can insert a credit card in an automatic teller 5,000 miles away from home and within

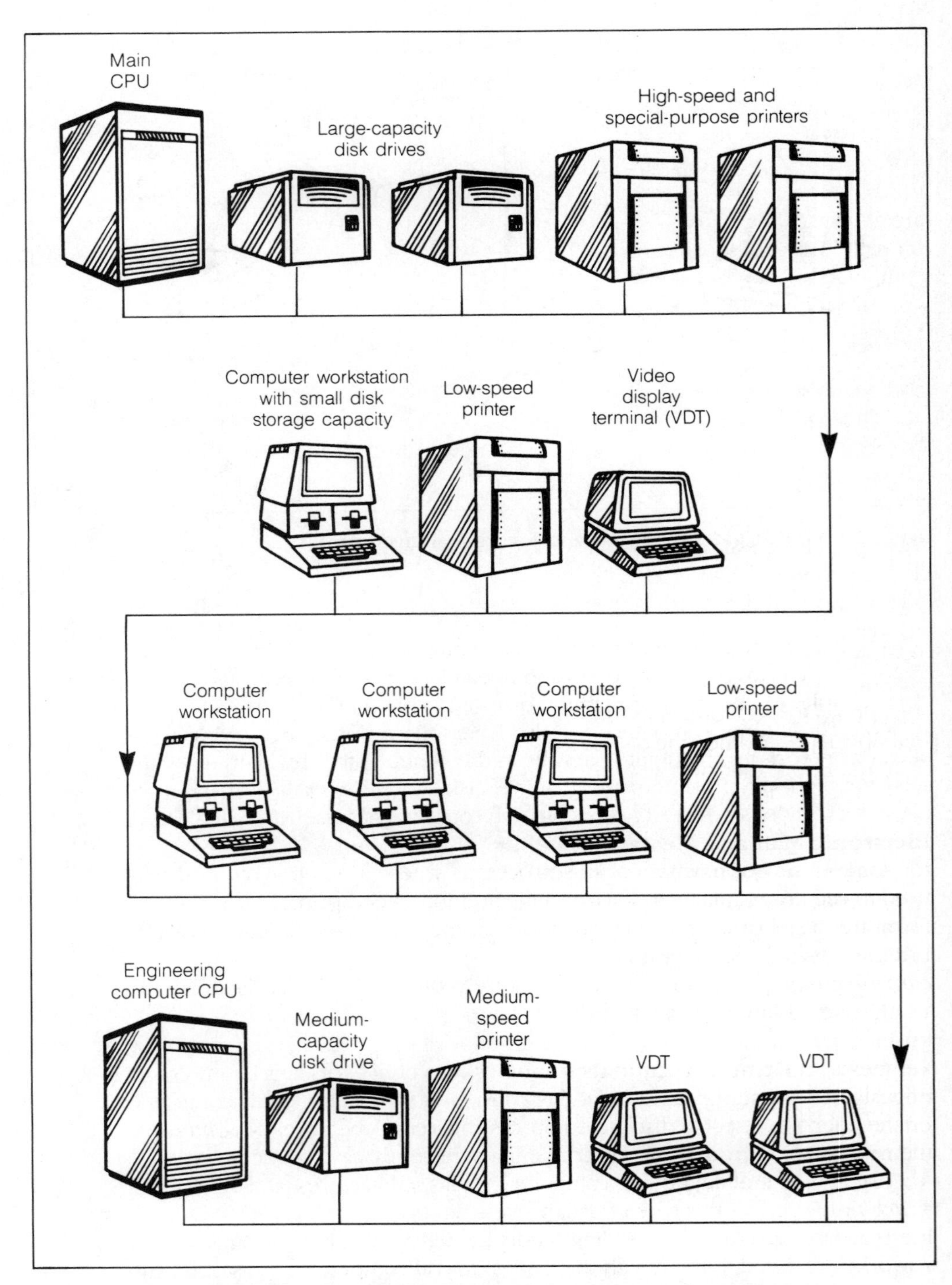

Figure 1.5. Data Communications Links Among Computers in Distributed Processing

seconds the bank's computer will verify the password, authorize the teller machine to issue cash, and debit the customer's account at the same time.

Many large and small retail stores are installing credit card checking machines that can read the credit card number from the magnetic strip on the card and transmit it to central-site computers, which authorize or reject the charge. Stores are also installing computerized cash registers, which can transmit information on sales and inventory on a daily (or even continuous) basis to central-site computers.

Airlines have installed self-service ticketing machines that allow the customer to insert a credit card and press a number of buttons to select his or her destination, flight, and even seat assignment. The central computer receives this information, processes it, and issues a ticket and boarding pass for the customer.

These modern applications require fast and accurate data communications. Errors in a customer's credit card number or in the amount of money to be dispensed at an automatic teller could be very expensive.

Newspaper and Magazine Delivery Newspapers and magazines are using data communications to speed up their delivery. They are transmitting the entire publication to remote printing facilities so that printing occurs in the local delivery area.

Some newspapers have even started to make their news stories and other information available for electronic delivery right to the customer's home. In this application, customers use their computer terminals to access, over ordinary dial-up telephone lines, the computerized data base that contains the information from the newspaper or magazine.

Electronic Mail and Message Systems Word processing has been with us for some time, but now word processors and data communications are being used to replace regular mail service. For example, after a letter is typed in final form, the word processor transmits the letter over telephone lines to a word processor at the receiving office miles away (see Figure 1-6). This is a form of *electronic mail.*

In one version of an *electronic message system,* the telephone operator of a business takes a call for a person who is busy on another call. The operator types the message into a computer terminal, and the message is transmitted instantaneously to a printer or computer terminal on the user's desk. Then the person on the phone can decide whether or not to put the current call on hold and take the new call.

A company's computer system can be programmed for *electronic message switching.* In this application, the computer user can type a short memo or message into the computer terminal on the user's desk, address that memo to a coworker, and the memo can appear on the coworker's desktop terminal within seconds (see Figure 1-6).

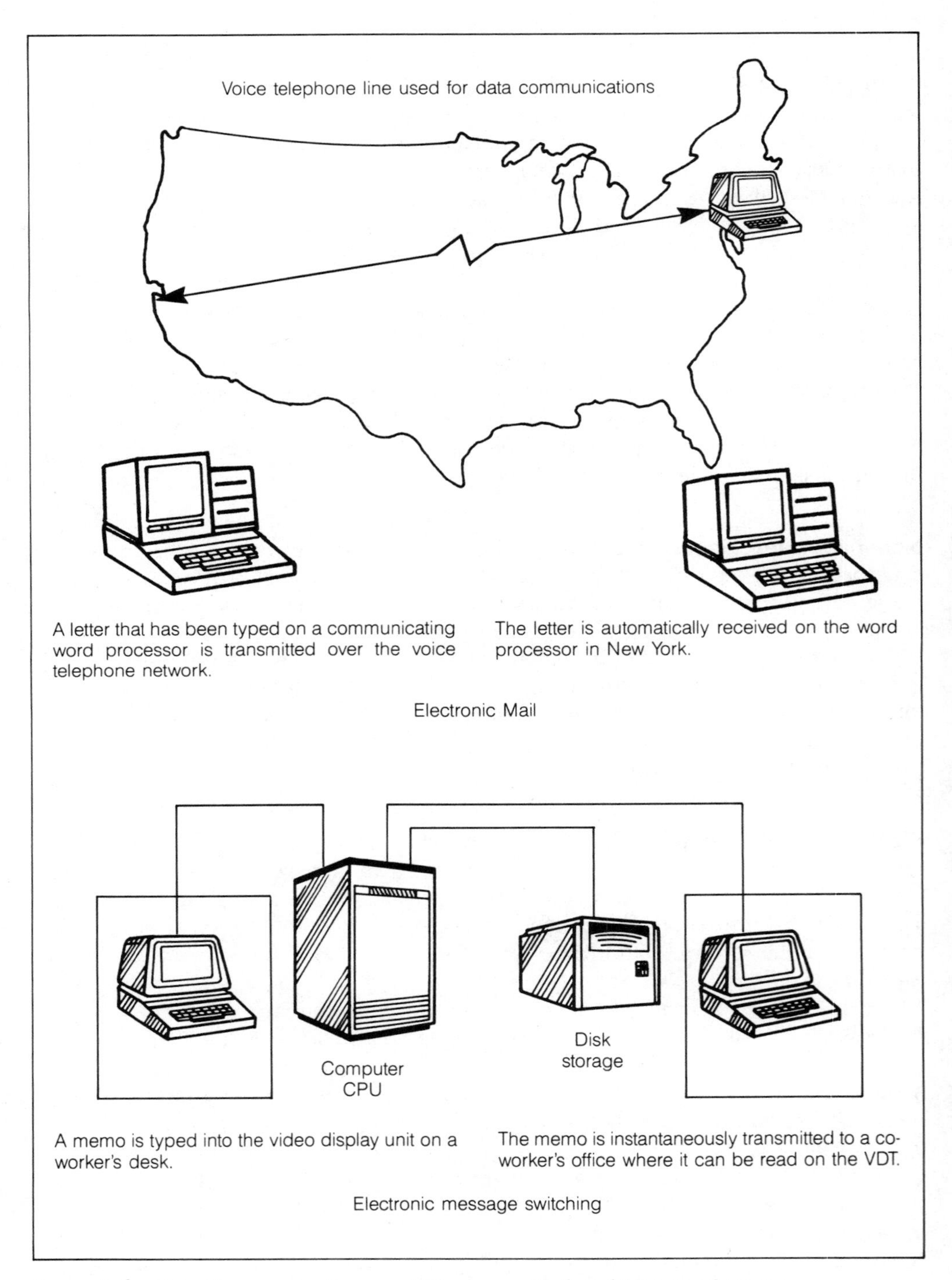

Figure 1-6. Data Communications and Electronic Mail and Message Systems

These are but some of the applications of electronic mail and message systems, which are growing at a very fast rate and which would not be possible without reliable data communications.

Store and Forward Systems These types of systems store a message while it is being created and then forward it later to its destination. They can be used for electronic mail, written messages, voice messages, telegrams, and so forth.

Central Data Banks A number of companies have developed computerized data banks that have a wide variety of personal and business information available to anyone with a computer terminal who has paid to join the data bank service. To use the service, you just dial the telephone number of the data bank, connect your computer or computer terminal to the telephone line, enter your identification number and possibly a security code, and then ask the data bank computer for the information you want. The data bank service transmits the desired information to your computer (or computer terminal) over the telephone line and bills you for the information later.

This application of data communications has developed largely because of the growing popularity of personal computers. With these small computers installed in more and more homes and offices, people have discovered how easy it is to call up over the telephone lines data bases that give them information on industry statistics, stocks and bonds, sports scores, want ads from the local newspapers, news stories, recipes, and more.

Fully Automated Dial-Up Data Communications Yet another fascinating application of data communications is the computer that automatically dials a telephone number and transmits information to another computer that has answered the call automatically. When the transmission has been completed, both computers then disconnect themselves from the telephone system and continue with other tasks. This approach can be applied to electronic mail, gathering sales information from remote stores, and a myriad of other applications.

There are many other examples, but these will give you an idea of the wide variety of data communications applications and a feel for what you can do with data communications as you use your own business or personal computer.

Data Communications Applications for the Future

As a result of dramatic cost decreases in electronics, over the next decade we can expect equally dramatic reductions in the cost of computers and data communications equipment. Microprocessors will become more powerful, less expensive, and much smaller. Memory will become so compact and cheap that hand-held computers costing less than a fancy dinner for two will have more memory than the standard mainframe computers of the 1960s. This will allow powerful computers to be used on a highly distributed basis and thus increase the need for data communications.

Store and Forward Switching Systems Many of the applications we have today are in their infancy in terms of the number of people that use them. For example, electronic mail and message systems exist today, as do some voice message systems, but these can be expected to grow explosively over the next ten to twenty years. The scenario at the beginning of this chapter provides examples of what to expect.

"Personal" Business Computers Both the use of distributed processing and the placement of computer terminals on the desks of managers and other professionals will continue to grow. As this growth brings more computer power to the users, these users will want access to more information contained in remote or centralized data bases, and thus the need for data communications will also continue to grow.

Home Computers Futurists say we will eventually have computer terminals in our homes over which we will receive much of our mail, newspapers, magazines, and other similar information. They say we will use our home computer terminals to shop for food, clothes, gifts, and even houses or cars.

Many people will be able to work at home using a computer terminal to communicate with their office. Computers will turn on and off the lights in our houses and businesses; control the heating and cooling systems; and keep us advised of the energy we are using, warning us when we exceed predefined limits.

Voice Recognition Scientists are predicting that voice-recognition technology will be sufficiently developed in the next decade for computer terminals to be manufactured that you can speak to instead of having to type information into. Once people can dictate letters directly to their computers, then electronic mail and message systems, with all of their data communications implications, can be expected to grow at fantastic rates. In the same way, computer terminals that can understand your spoken request to locate information in a letter or memo that was received months ago will dramatically increase the need for data communications.

POINTS TO REMEMBER

All of the applications mentioned in this last section will probably become widespread reality in the next decade, and they will all contribute to what will become an explosion in the use of data communications. As the home and the business world become more computerized, data communications will become an increasingly important aspect of all of our lives. So, in the following chapters you will learn the vocabulary of data communications, come to grips with basic data communications concepts, and be shown how to implement the data communications applications that you need.

2

HOW VOICE AND DATA GET TO WHERE YOU NEED THEM TO GO—ANALOG AND DIGITAL TRANSMISSION

In Chapter 1, we looked at the way people confuse voice communications and data communications. Basically, people tend to identify a certain type of original information (or *signal*) with a certain transmission technique (or *mode*); voice communications is identified with analog transmission, and data communications (or computer-generated information) with digital transmission. This chapter explains why these associations are incorrect, and it describes how both voice and data are transmitted in either digital or analog mode and how data communications uses the voice telephone network. Thus, this chapter gives you a picture of the true inner workings of communications technology.

ANALOG TRANSMISSION AND THE TELEPHONE SYSTEM

The telephone system is the most familiar example both of voice communications and of analog transmission. Two people, in the same building or thousands of miles apart, can communicate by talking in a normal voice into a telephone (see Figure 2-1). The mouthpiece (or transmitter) of the telephone converts the sound waves of the speaker's voice into electrical signals that vary proportionately to the frequency and loudness of the speaker's voice. These electrical signals are called "analog signals" because the frequency and voltage (or level) of the electrical signal are used to represent—that is, are an *analog* of—the frequency and loudness (or level) of the original sound wave. These analog electrical signals are transmitted to the telephone at the other end of the connection, and there they are converted back into sound waves that the listener hears through the receiver. Thus, the telephone system is referred to as an analog transmission system.

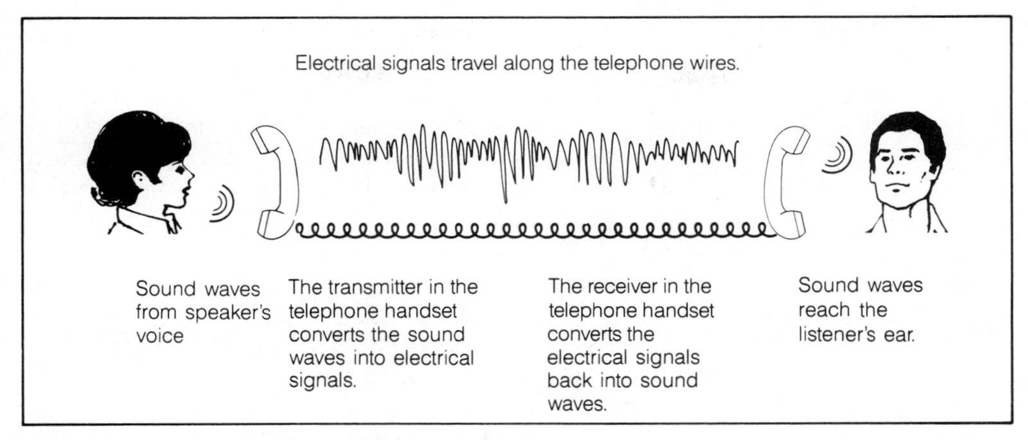

Figure 2-1. Analog Transmission Using the Telephone System

DIGITAL TRANSMISSION AND COMPUTER COMMUNICATIONS

Just as people need to communicate directly with one another in order to exchange or transmit information, such as "What's for dinner?," "How much does it cost?," or "The kids are fine, don't worry," so it is also necessary for machines such as computers and terminals to communicate—that is, to transmit data—to one another.

When a computer terminal generates a message for another terminal or a computer, it generates electrical signals that represent an "on" or "off" condition, represented by 1's or 0's. This "language" of computers is called a binary system because it consists of only two symbols (0 and 1) or two electrical states ("off" and "on"). And, as we explained in Chapter 1, because the binary language consists of only the two binary digits 0 and 1 (also called "bits"), it is also referred to as a digital language.

Computers, terminals, and other similar devices commonly use this digital language to transmit information between them when they are located within a few feet of each other. Remember that the transmission of information or data in digital form (that is, in 0's and 1's) is called digital transmission; thus, as we mentioned in Chapter 1, the terms data communications and digital transmission are frequently used interchangeably.

However, digital transmission does not always mean data communications, and vice versa. In the balance of this chapter we will explain how most long-distance voice communications actually involves digital transmission and how most long-distance data communications actually involves analog transmission.

In other words, it is not the form the signals are in during transmission (that is, analog or digital) that distinguishes voice communications from data communications. Both voice and data can be transmitted by either digital signals or analog signals. With the definitions we are using, whether a signal is classified as data communications or voice communications depends solely upon its source and destination (that is, whether it goes from machine to machine or from human to human) and its form at the source (that is, whether it is machine language or a human voice).

A TECHNICAL LOOK AT HOW ANALOG AND DIGITAL TRANSMISSION WORK

You talk into the telephone in your office or type into your computer terminal that's connected to your company's telephone system. You expect that voice or data conversation to be transmitted through the telephone system to the telephone or computer at the other end of the line. Some form of analog or digital transmission technique must be used to get that voice or data conversation from your office to the other end of the line. In this section, we examine some of the various analog and digital transmission techniques that can be used.

To show you exactly how analog and digital transmission work, let's look at how a voice signal can be transmitted by both analog and digital means. This section may, at first, seem "too technical," and, in fact, it is the most technical section in the entire book. But the method of transmission is a very important part of data communications, so keep working at understanding this information until you feel you've grasped the basic concepts.

Continuous Analog Transmission

We have described how a voice signal is converted into an analog electrical signal (look again at Figure 2-1), but how is that signal transmitted through the telephone system on a local telephone call? A constant voltage (supplied at the telephone company central office) is applied to a pair of wires that run between the telephone company central office and a home or business (this pair of wires is called a "local loop"). The pressure of the sound waves against the transmitter or microphone causes this constant voltage to be modulated (that is, to be varied) so that it represents, or becomes the analog of, the sound wave. The level of the electrical voltage of the modulated signal varies directly with the level of the sound wave itself, and the frequency of the modulated signal varies directly with the frequency of the sound wave.

For a local call, the local loop is connected at the central office to another local loop; that is, it connects to the loop between the central office and the home or business receiving the call. The analog electrical signal travels along the second local loop until it reaches the receiver or speaker; there the signal is changed back into sound waves, which enable the user to "hear" the person talking at the other end. This is called *continuous analog transmission;* it is the simplest form of analog transmission and also the simplest form of voice communications through the telephone system.

Sampling

In addition to the continuous form of analog transmission, other forms exist based on a technique called *sampling.* If you plot a sound wave on graph paper, with the level on the vertical axis and time on the horizontal axis, you can see how sampling works (see Figure 2-2). For every square on the horizontal axis (every time interval), place a dot where the wave form crosses an imaginary vertical line that extends through the center of the square. You are creating a series of dots that lie on the wave form and are equally spaced in terms of the horizontal measure of time (see Figure 2-3); these are called the *sampling points.*

It is easy to see that connecting these dots together by straight lines produces a figure that very closely resembles (or is the analog of) the original wave form (Figure 2-4). The smaller the squares—that is, the more frequent the sampling interval—the closer the dot-and-line figure approximates the original wave form (Figure 2-5).

In other words, this sampling technique determines a wave form by looking at the wave in the middle of each short and equal sampling interval and then determining the level of the wave at that specific point, called the sampling point. Figure 2-6 shows another example of this technique with the level of the wave form expressed by a number between 0 and 5.

Transmission of Sampled Signals

Once the wave has been sampled and the height has been determined at each sampling point, how can these data (the level at each sampling point) be transmitted? There are both analog and digital methods for transmitting these data; the following subsections describe two analog methods and one digital method. You will find all three methods mentioned frequently in data communications literature.

Pulse Amplitude Modulation One of the most popular analog transmission techniques frequently used in private branch exchanges (PBX's) is called pulse amplitude modulation (PAM). A PBX is an automated telephone switching system used by many businesses, which is characterized by the use of a central operator and by the user dialing 9 to get an outside line.

With pulse amplitude modulation (PAM), a series of equally spaced pulses are transmitted over a pair of wires (or other transmission medium). The level (or amplitude) of each of these pulses is modulated or varied so that it is directly proportional to the level of the wave form at the sample point.

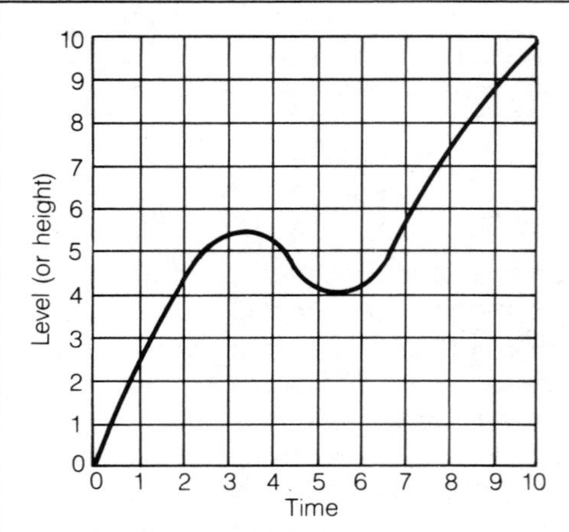

Figure 2-2. Understanding Sampling—Step 1: A Sound Wave Plotted on a Time Chart

The units of measure in this graph and in the ones that follow could be volts (along the vertical axis) and milliseconds (along the horizontal axis), but these graphs have purposely been kept abstract so that you can grasp the general concept of sampling rather than be distracted by specific quantities or examples.

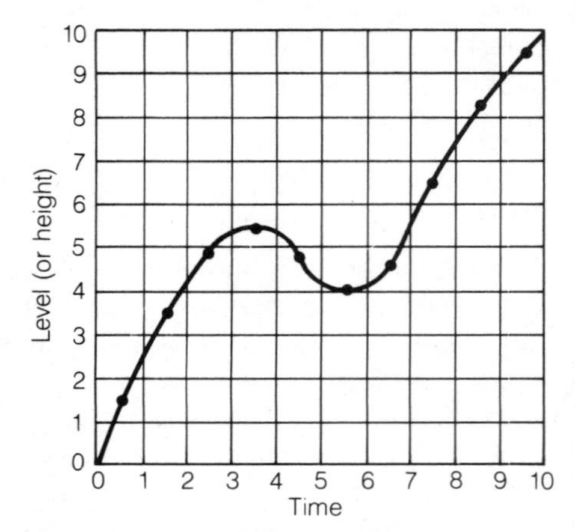

Figure 2-3. Understanding Sampling—Step 2: Find the Sampling Points at the Center of Each Time Interval

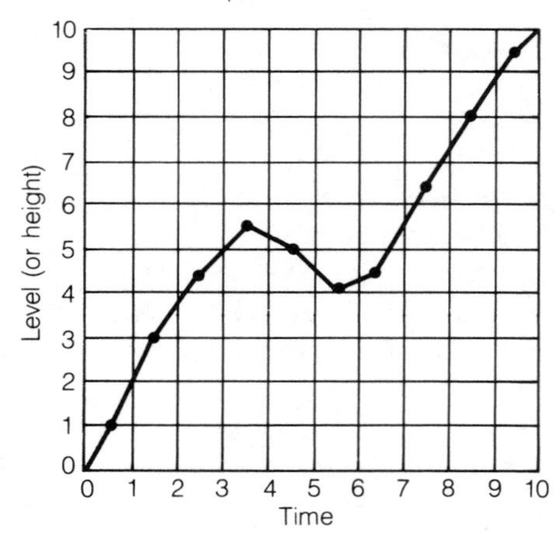

Figure 2-4. Understanding Sampling—Step 3: Sampling Points Connected by Straight Lines

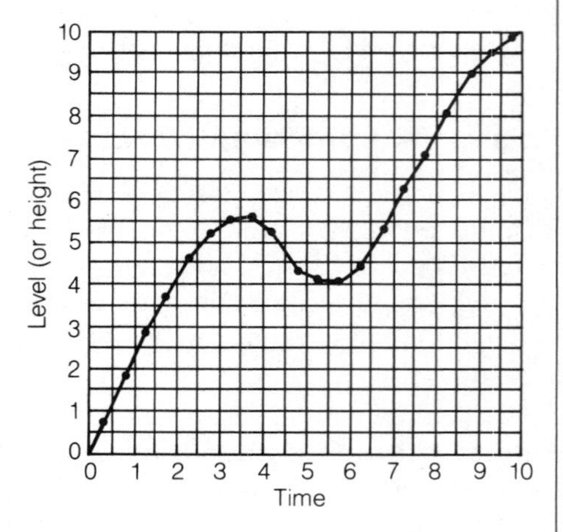

Figure 2-5. Understanding Sampling—Step 4: If More Frequent Sampling Points are Connected by Straight Lines the Result More Closely Approximates the Original Wave Form in Figure 2-2

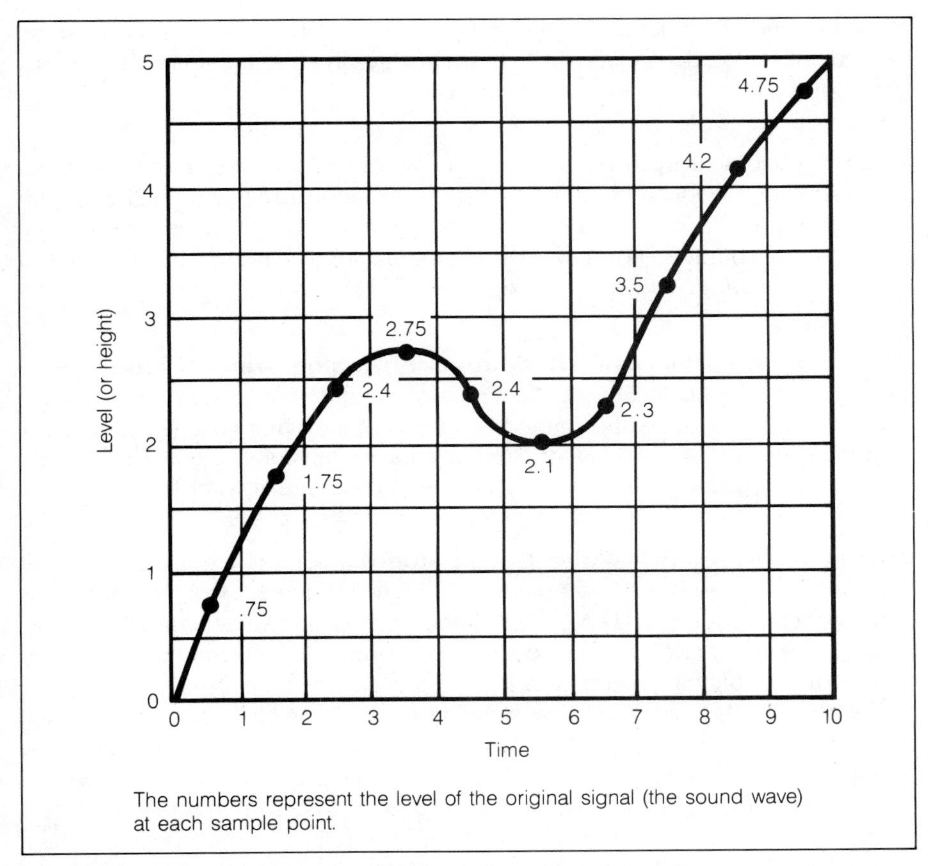

Figure 2-6. Understanding Sampling—Step 5: Using Numerical Values with Sample Points

Let's suppose that the wave form is sampled every 1/1,000th of a second. The level of the wave form at the first sample point is 2.15; the level at the second sample point is 2.30; at the third, 2.65; the fourth, 3.10; and the fifth, 3.50. Let us assume that the pair of wires would normally have a voltage across them of 0 volts. With this analog transmission technique, every 1/1,000th of a second the voltage on the pair of wires would be changed, or modulated, to be equal to the level of the sample. Thus, for the first 1/1,000th of a second, the voltage would be 2.15 volts; for the next 1/1,000th of a second, the voltage would be changed to 2.30 volts; then to 2.65 volts; then to 3.10 volts; then to 3.50 volts. Thus we can see from Figure 2-7 how a series of pulses, or voltage changes of equal duration, create the analog of the wave form we want to transmit.

At the other end of the circuit (that is, at the other end of the pair of wires), the series of pulses are converted back into a representation of the original wave form. Depending upon the sampling rate and the use of a circuit that can generate curved lines instead of straight lines in between each sampling point, the original wave form can be very closely reproduced.

Pulse Width Modulation Another analog transmission technique is known as pulse width modulation (PWM) or pulse duration modulation (PDM). With this technique, the level (or amplitude) of the transmitted pulse is constant, but the width (or duration) of the pulse is modulated, or varied, with the level of each sample point on the original signal (see Figure 2-8).

The Common Aspect of All Methods of Analog Transmission Other analog transmission techniques are possible, but one common characteristic always exists. A signal is transmitted by means of a varying voltage (or possibly a varying current), which is modulated in some dimension (such as level, width, frequency, or phase) so that it is the analog of the original wave form.

Digital Transmission—Pulse Coded Modulation There are also digital techniques for transmitting sample levels; the technique we will look at here is pulse coded modulation (PCM). (Remember that we are still talking about transmitting a sound wave, so we are still discussing voice communications even though it uses digital transmission.)

With PCM, the wave form is sampled just as with the analog techniques, but, instead of representing the level of the sound wave at each sample point by modulating some dimension of a pulse, pulse coded modulation involves *encoding* the level at each sample point; that is, the level is given a numerical value, often a whole number or integer. For example, we can superimpose a new scale of whole numbers on the graph in Figure 2-6 (see Figure 2-9), and then we can encode the level of the wave at each sample point as the integer closest to its level.

Once the level at each sample point is encoded as an integer, then that number is translated into a binary number—that is, a series of 0's and 1's (see Figure 2-10). This binary-coded number is transmitted as a series of pulses of uniform width, that may have a level that represents either the 0 or the 1 (see Figure 2-11). At the other end of the circuit, the pulse levels (or voltages) are reinterpreted back into 0's and 1's and are used to recreate the wave form.

The binary number system is very easy to transmit electrically because the digit 0 can be represented by a zero voltage and the digit 1 can be represented by a voltage of one volt or some other voltage such as ten volts. In the situation where zero and ten volts are used to represent the binary digits 0 and 1, the receiving device need only determine if there are approximately zero volts on the pair of wires or something close to ten volts. If it sees nine volts, that's good enough to be interpreted as ten volts, or a 1.

Pulse coded modulation is used in most of the long-distance telephone conversations in today's telephone system. Because this transmission technique

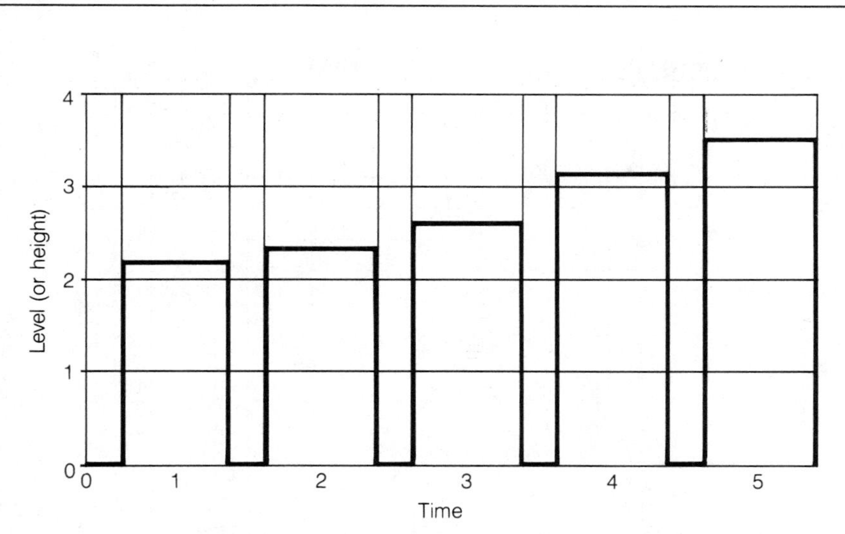

The voltage on the transmission medium (such as the telephone line) is modulated to match the level of the original sound wave at each sampling point.

Figure 2-7. Transmitting Sampled Signals: Pulse Amplitude Modulation

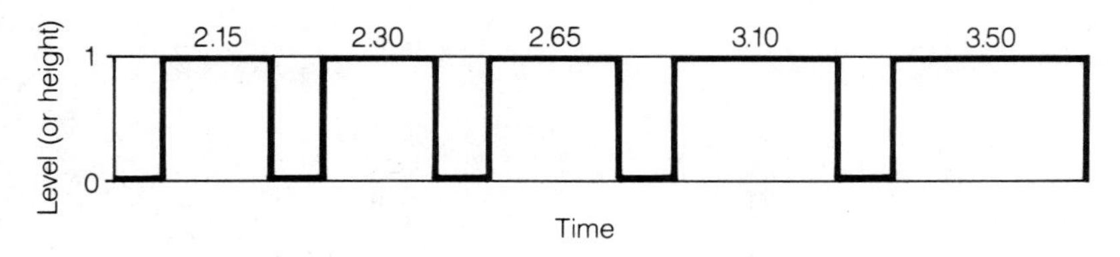

With this analog transmission technique, the amplitude of the pulse on the telephone line remains the same, but the duration of each pulse varies to represent the amplitude of the original sound wave at the sampling points.

Figure 2-8. Transmitting Sampling Signals: Pulse Width Modulation

translates the voice signals into binary numbers, which are the same type of numbers processed by computers, some people call this data communications. But we prefer to consider it voice communications because the information (or data) originated as a human voice.

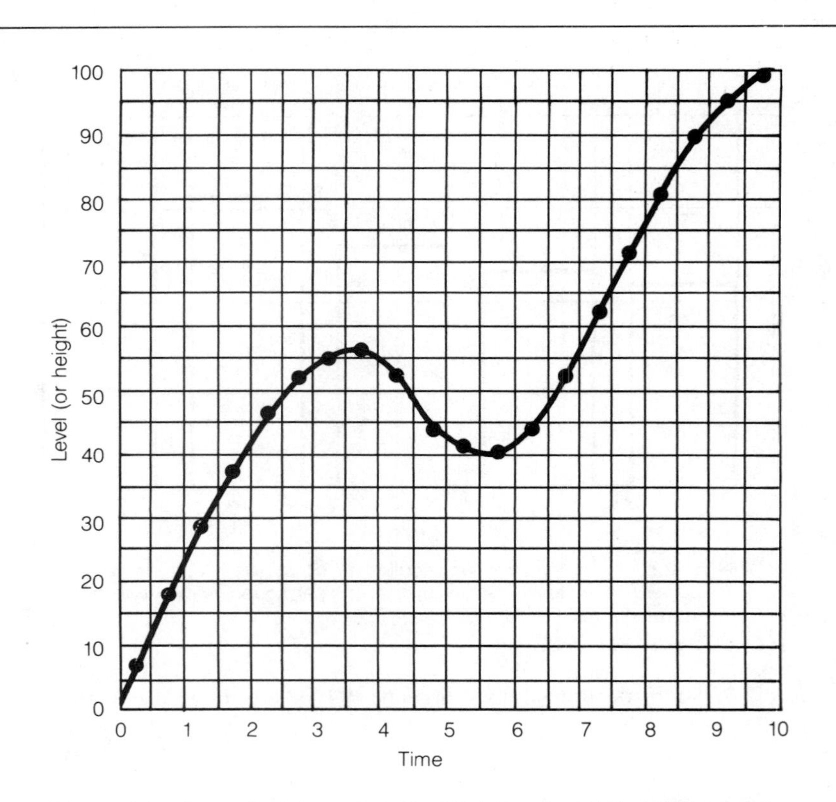

Figure 2-9. Encoding Sample Point Heights for Digital Transmission

7	0000	0111	42	0010	1010
18	0001	0010	41	0010	1001
28	0001	1100	44	0010	1100
37	0010	0101	53	0011	0101
46	0010	1110	68	0100	0100
52	0011	0011	71	0100	0111
55	0011	0111	81	0101	0001
56	0011	1000	90	0101	1010
53	0011	0101	95	0101	1111
43	0010	1011	99	0110	0011

Figure 2-10. Encoding Sample Point Heights into Binary for Digital Transmission

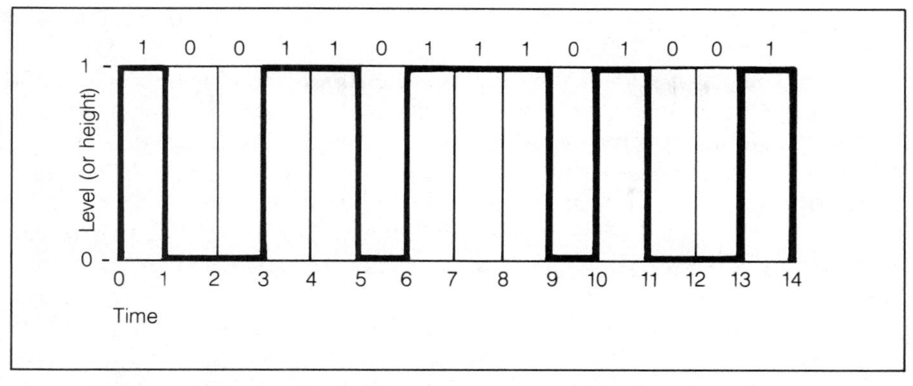

Figure 2-11. Pulse Coded Modulation: Transmitting Sample Point Heights in Binary Code

You can now see why the line dividing voice and data communications is a little blurred. As we go through this and other examples, keep in mind the distinction between the *mode* of the transmission (whether it be analog or digital) and the *signal* (whether it be voice or data).

Why PCM Digital Transmission Is Superior to Analog Transmission

We can understand why pulse coded modulation is considered superior to analog techniques over long distances by seeing what happens to an electrical pulse as it travels along a pair of wires. As the pulse travels down the wires, it meets resistance from the wire itself. In fighting this resistance, the pulse loses strength—that is, its voltage or level is reduced. The pulse is also subject to a number of forces that tend to alter its shape and duration as well as its level. The longer the distance the pulse travels, the more severe the attenuation (the lowering of its voltage) becomes, until eventually the pulse could disappear or become undetectable at the other end.

To overcome this undesirable electrical phenomenon, the telephone system uses amplifiers at periodic points along the wire to amplify or strengthen the pulses. But, because the pulses have become distorted and their level, width, and shape have changed from the original pulse, there is no way for the amplifier exactly to reproduce the original pulse if it is an analog pulse. If the amplifier sees an incoming pulse of 3.5 volts and knows that it has been attenuated, should the outgoing pulse be 4.0 volts, 4.05 volts, or 4.10 volts? In an analog system there are an infinite number of levels or widths that the pulse could be. Also, the attenuation introduced by the system varies, depending upon things like the temperature. The amplifier obviously introduces some inaccuracy in trying to recreate the original pulse because it cannot tell exactly how much attenuation has taken place.

With a digital system, however, the pulses are always of the same width, always of the same shape, and always exactly of a level representing a 0 or a 1. Although different systems may use different voltages to represent 0's and 1's, the same unique voltages will always be used in any one system. Thus, in a system that uses zero volts to represent a 0 and ten volts to represent a 1, if a repeater or regenerator (which is the equipment used in a digital system instead of an amplifier) sees a badly misshapen pulse of five, six, seven, or eight volts, it knows it must regenerate a perfectly shaped pulse of ten volts; thus it exactly reproduces the original pulse without introducing any error (see Figure 2-12). It is this ability to regenerate the signal on a relatively error-free basis that makes a digital system superior to an analog system for long-distance transmission and in many other situations.

Errors in Digital Transmission

We will discuss two types of errors that can occur with digital transmission so that you will not get the idea that digital transmission is completely error free.

Digital Error Noise, or random electrical signals, can be induced onto a transmission line by several means, such as the signals from other wires in the same cable or from the electromagnetic radiation created during an electrical storm. If the noise on the line causes a pulse of zero height to reach a high enough voltage to be interpreted by the regenerator as a 1, an error results. Conversely, if a pulse of 1 is attenuated enough to be interpreted as a pulse of 0, again an error results. This type of error is called a *digital error*.

Encoding Error When the wave is being encoded, some error can occur due to the encoding process. For example, a wave form with sample point heights of 4.1, 4.2, 4.3, 4.4, 4.3, 4.2, and 4.1 would be encoded as 4, 4, 4, 4, 4, 4, 4 and would be reproduced as a straight line. The up-and-down motion of the original wave form, although quite minor, would have been lost. This is called an *encoding error* or a *quantization* error.

Summing Up Transmission Techniques in the Telephone System

We have now described how analog and digital transmission techniques work, and we have seen how our telephone system, which is designed as a voice communications system, uses both analog and digital techniques for transmitting

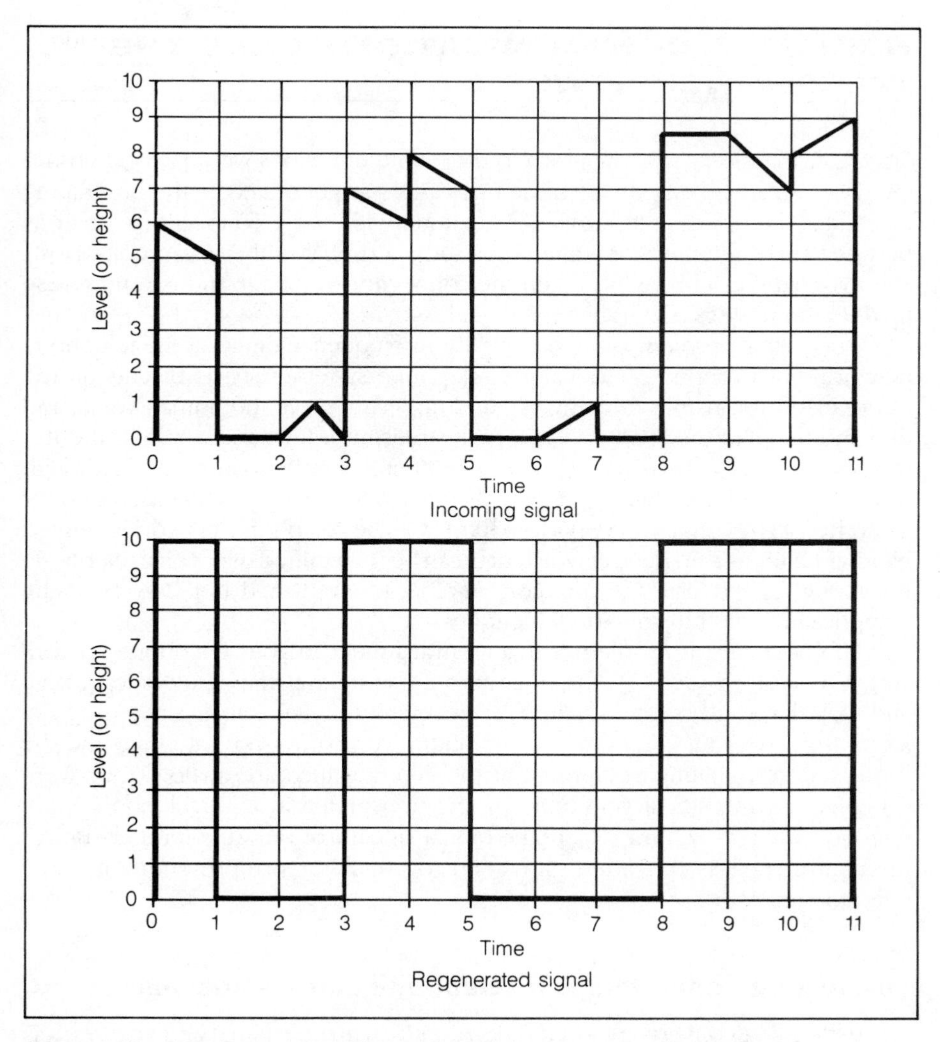

Figure 2-12. Misshapen Digital Signals Can Be Regenerated Accurately

voice signals. The continuous analog technique is used on local loops (from homes and businesses to the telephone company central office) and some medium- and long-distance routes as well. The digital sampling technique (pulse coded modulation) is used on most long-distance routes. We have also looked at the analog sampling techiques pulse amplitude modulation and pulse width modulation, which are used in some computerized PBXs.

HOW DATA COMMUNICATIONS USES THE TELEPHONE NETWORK

Obviously the most direct approach to data communications would be to transmit the data digitally in exactly the same form as it was generated by the originating machine. This is actually done in situations with short connections, such as between two computers or a computer and a terminal in the same room. However, you run into problems when the connection is longer and you try to use the telephone system to transmit a digital signal.

One major problem is the bandwidth or frequency range of the telephone network. For economic reasons, the telephone system restricts or cuts off the frequencies it transmits at the lower and upper range of the human voice, and thus the telephone system is designed to transmit information only in the frequency range of approximately 300 to 3300 cycles per second (cps), although its total bandwidth is slightly wider. (The term *hertz,* or *Hz,* is also used to refer to cycles per second, as in 300 to 3300 Hz.) The telephone network's limited bandwidth limits the speed at which data can be transmitted over voice telephone lines. High-speed data transmission *can* be accomplished but only by using complicated data transmission techniques.

The other major problem is that the transmitter in your telephone set converts the sound waves of your voice into a continuous analog electrical signal and that almost all computers initially generate a digital signal. Since the local loops from your telephone to the telephone company's central office and the local loop terminating equipment at the central office are analog (they were designed for the signal generated by the transmitter in your telephone), the telephone system will not accept the digital signal from most computers unless the digital signal is converted into an analog signal before presenting it to the telephone network.

Connecting Computers and Telephone Lines—Modems

The device that converts the digital signal into an analog signal and vice versa is called a *data set* or *modem* (the abbreviation for *mo*dulator/*dem*odulator). Modems will be described in detail in Chapter 5, but you need to understand here how they provide an interface between computers and the telephone system.

If the data are being transmitted by means of a long-distance telephone call, then the analog electrical signal generated by the modem is probably sampled, coded, and transmitted (by the telephone company's equipment) as digital 0's and 1's until it reaches the other end of the connection. There the digital signal is reconverted to an analog signal, which is presented to the modem. The modem then reconverts the analog signal back to a digital signal before passing it on to the receiving machine (see Figure 2-13).

Note that, although composed of 0's and 1's, the digital signal created by the

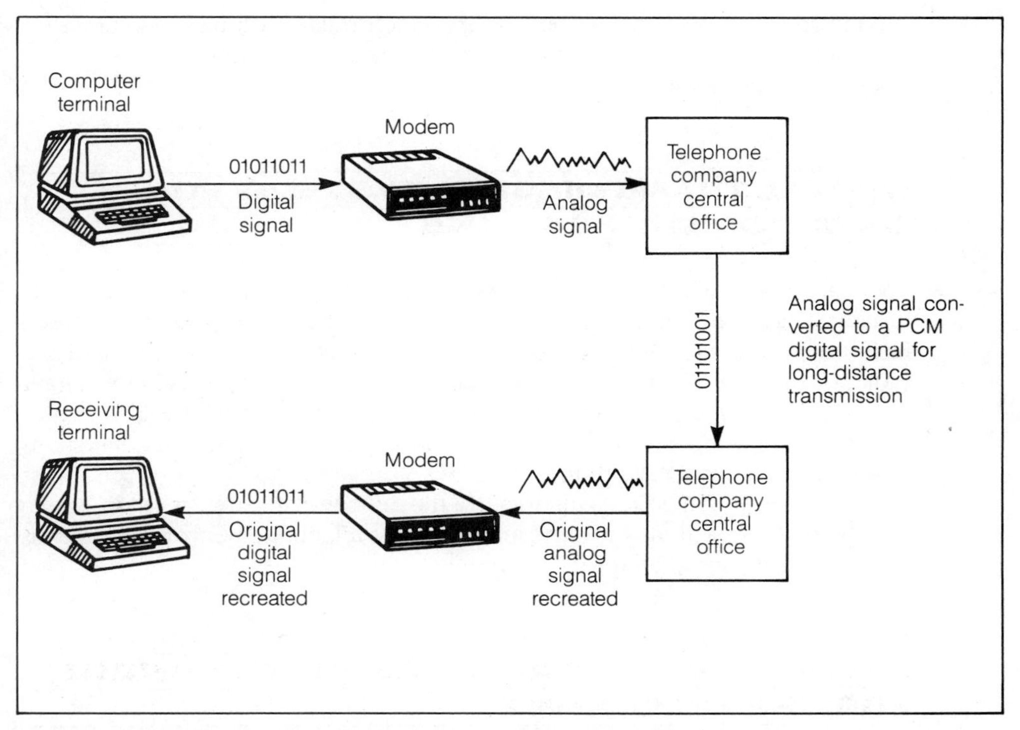

Figure 2-13. Data Communications on the Telephone Network: Translating Digital Signals to Analog and Back Again

telephone company in the long-distance part of the transmission consists of a different series of 0's and 1's than the original digital signal generated by the computer. This is because the original digital signal is coded by the modem as an analog signal, and it is this analog signal that is sampled and coded by the telephone company for long-distance digital transmission. Nevertheless, when these coding processes are reversed at the receiving end, the digital signal presented to the receiving machine will be identical to the digital signal generated by the originating computer.

Problems with Transmitting Data over the Voice Network

The first problem with using the voice network for data communications is that the speed at which data can be transmitted is limited, as we mentioned before. Second, modems can be very expensive, especially as the speed of data transmission increases. Third, the telephone system—with all of the necessary signal conversions and equipment built into the network for voice communications—

can introduce errors. Finally, you may encounter relatively long set-up times for data calls. Thus, the existing voice communications telephone network is not optimal for data communications.

DIGITAL TRANSMISSION SYSTEMS FOR DATA COMMUNICATIONS

There are all-digital transmission systems available for short-, medium-, and long-distance data communications. Some use satellites and some use land-based facilities such as metallic or copper wires, coaxial cables, optical fibers, digital radio, or microwave. These eliminate the need for modems, provide for higher data speeds, and solve other problems you encounter when using the voice telephone network for data communications. Digital communications systems are discussed in more depth in Chapter 9.

However, you must keep in mind that the most available transmission system and the one you will most likely use is the voice telephone network, which was not specifically designed for data communications.

A NEW DIGITAL TRANSMISSION TECHNIQUE FOR THE TELEPHONE

A new electronic development provides one final illustration of the blurring of the line between voice and data communications: CODECs are electronic circuits that code and decode analog voice signals into digital signals. Until recently, CODECs have been part of a PBX, which, in turn, was hooked up to the telephone system. But now, because technological advances are bringing the cost of these semiconductor chips down to a very low price, it will be possible to install the CODEC directly into the telephone set itself, and then the telephone will be able to convert the human voice directly into a digital signal. As a result, digitized voice signals will be transmitted like digital computer signals, and it will be possible to transmit both voice and data digitally on the same wires at the same time. In this environment, most of the practical distinctions between voice and data communications disappear.

We will see CODECs used directly in the telephone set first with digital PBXs such as Rolm Corporation's CBX and AT&T's System 85. When these PBXs use a telephone with a built-in CODEC and are connected to a data terminal or computer, they can transmit both voice and data simultaneously over one or two pairs of wires by using digital transmission techniques.

Eventually, we will see this approach expanded to the local loops from the telephone company central office, so that all phones will be "digital," and you won't need a PBX in order to use digital transmission. However, because of the large investment that telephone companies have made in analog local loops and central offices, it will be many years before digital telephones will replace today's analog telephones for direct connections to the telephone company's local office.

INTEGRATED SERVICES DIGITAL NETWORKS—TRANSMISSION OF THE FUTURE

Plans are now being made, both in the United States and abroad, for a new generation of communications systems called *integrated services digital networks* or ISDN. An ISDN will enable the user to transmit voice, data, video, and any other type of communications that is digitally encoded, thus enjoying the speed, economies, and accuracy that a fully integrated all-digital communications system can provide.

POINTS TO REMEMBER

In this chapter, we have defined both analog and digital transmission and looked at the specific transmission techniques pulse amplitude modulation (PAM), pulse width modulation (PWM), and pulse coded modulation (PCM). We have also seen how either analog or digital transmission can be used for either voice or data communications. As a result, you should realize that the *form* during transmission (whether analog or digital) is not what determines whether the process is voice or data communications.

3

DATA COMMUNICATIONS BASICS—THE DEFINITIONS YOU NEED

This chapter is the one you would like to skip, but please don't! Here we define and discuss the technical words used in data communications. Unfortunately, you cannot understand data communications without knowing what a bit is, what the difference between synchronous and asynchronous transmission is, or what an RS-232C interface is, to name just a few examples. So, this chapter gives you definitions for the various terms you will come across frequently in both your readings and discussions on data communications.

THE COMPUTER

A device we generally refer to as a computer consists of a central processing unit (CPU), memory, a power supply, and an operating system. The operating system is the software that manages the overall use of the computer and that allows a user to interface with the computer without having to write every instruction in machine language—that is, in a series of 1's and 0's.

THE CRT TERMINAL OR VIDEO DISPLAY TERMINAL

The computer is, however, relatively useless by itself. In order to use it effectively, you need to be able to communicate with it. A common means of communicating with the computer is through a CRT (cathode ray tube) terminal with a keyboard. This terminal with a CRT screen and keyboard is often referred to as a "computer terminal" or simply as a "CRT," although CRT refers to the tube or screen part of the terminal only. The CRT terminal is also referred to as a video display terminal (VDT) or video display unit (VDU).

DATA STORAGE AND PRINTING CAPABILITY

If you need to store more data than the computer's internal memory can hold (as most users do), then you will add a disk or tape drive to your system for data storage. Most users also want the capability of having the computer print the data collected and processed; if you do, you will also need to connect a printer to the computer.

PERIPHERAL DEVICES

Devices that connect to the computer—such as a CRT terminal, disk drives, tape drives, and printers—are all referred to as "peripheral equipment," "peripheral devices," or simply "peripherals." It is, of course, necessary for the computer to be able to talk to or communicate with its peripherals in order to use them; this is where data communications comes in.

DATA TERMINAL EQUIPMENT AND DATA COMMUNICATIONS EQUIPMENT

Data communications jargon is replete with abbreviations and acronyms. Two acronyms you will come across frequently are DTE and DCE.

Data Terminal Equipment (DTE)

Data terminal equipment (DTE) is sometimes also referred to as data processing terminal equipment (DPTE). DTE refers to all data processing equipment and other sources of data such as computers and communications controllers and to all peripheral equipment such as computer terminals and printers.

Prior to deregulation, which effectively occurred January 1, 1979, as a result of the FCC's 1968 Carterfone decision (see Chapter 11), all telephone services and the related data communications equipment (DCE) such as modems were supplied by the local telephone company or, in the case of long-distance service, by AT&T on a "regulated" monopoly basis. Therefore, prior to 1979, DTE simply referred to all data processing and communications equipment supplied by the data processing hardware suppliers or manufacturers to the user.

Data Communications Equipment (DCE)

Data communications equipment (DCE) or data circuit-terminating equipment, refers to all communications equipment (for the most part modems or their functional equivalents) that connects to the telephone line side of data terminal equipment (see Figure 3-1).

The term DCE originated prior to deregulation and simply referred to all data communications equipment supplied by the telephone company; it therefore included modems or data sets. Today (that is, after deregulation) data processing equipment manufacturers make modems, yet modems are still classified as DCE. In fact, modems along with multiplexors (defined in Chapter 8) are the two pieces of communications equipment that are usually thought of when the term DCE is used.

CONNECTING DATA TERMINAL EQUIPMENT FOR DATA TRANSMISSION

The terms circuits, lines, and links all usually refer to the physical communications media over which data are transmitted. The physical media can be ordinary twisted-pair telephone wire, coaxial cable, microwave, satellite radio, fiber optics, or any other communications medium used to carry the data. The different transmission media available are discussed in Chapter 9. In this section we will present definitions of the "generic terms" that describe the following physical connections: circuits, lines and links.

Circuits

The term *circuit* is the most general of all three terms and is used to refer to all types of physical communications media. The word circuit is also occasionally used to refer not to the physical media itself but rather to one of many two-way communications paths that are derived from a single physical communications media. For example, we can say, "We have derived three data circuits from one voice-grade telephone line," which means that three simultaneous two-way data communications conversations can be carried on one physical circuit (the telephone line).

Point-to-Point Circuits A point-to-point circuit is a circuit with only two connections—that is, the data terminal equipment is attached at only two points on the circuit, one at each end.

Multipoint Circuits A multipoint circuit is a circuit with DTE connected at intermediate points in addition to the two end points.

Line

The term *line* is usually used to refer to a wire circuit, or a circuit obtained from a common carrier such as AT&T or the local telephone company. This can be

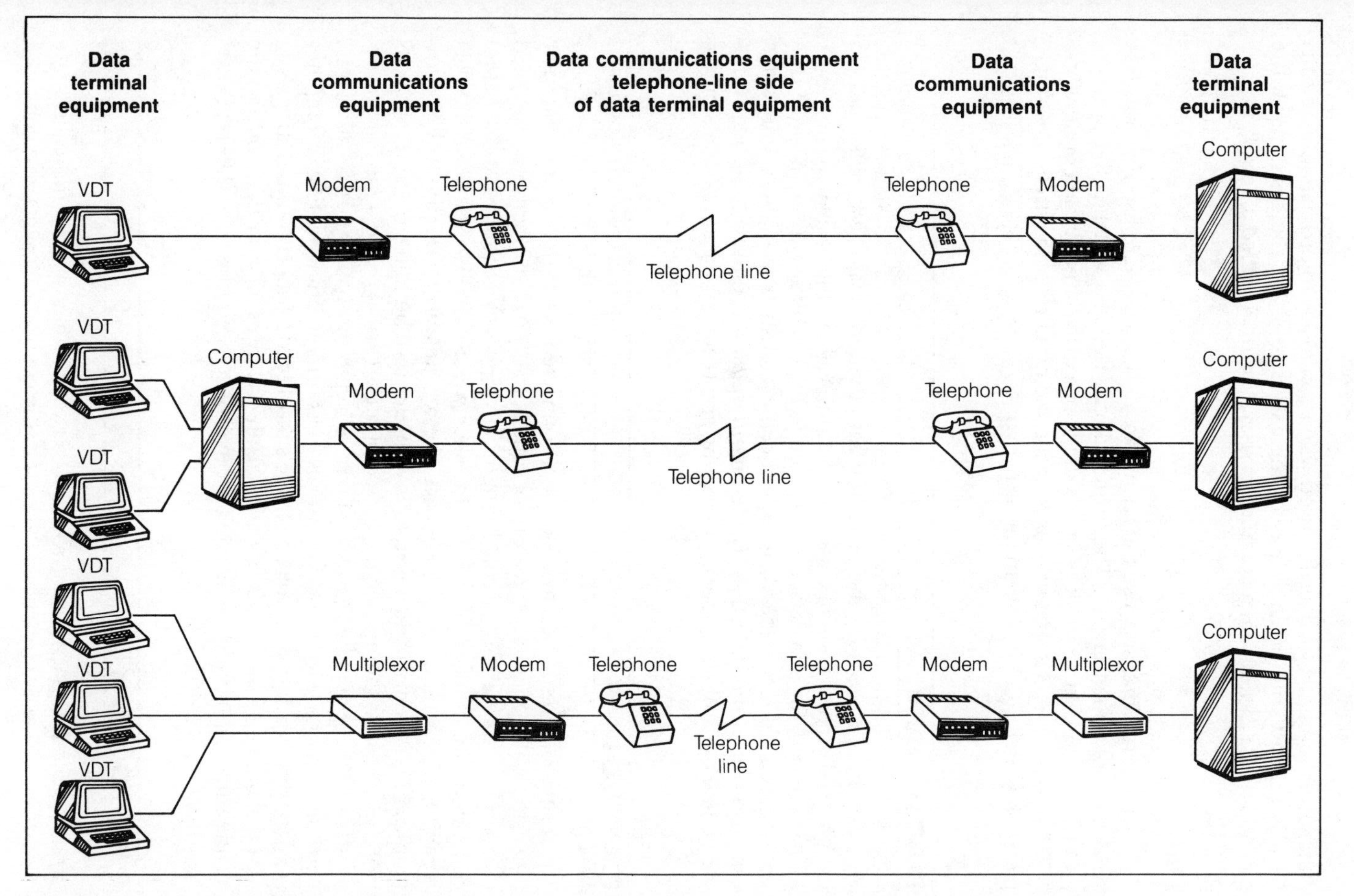

Figure 3-1. How Data Terminal Equipment and Data Communications Equipment Relate to One Another.

either a dial-up line (such as that used when you make an ordinary phone call) or a "dedicated" or "private" line for the exclusive use of the customer. If the private line is leased from a common carrier, it is often called a *leased line*.

Link

Finally the term *link,* and also the term "circuit segment," are used to refer to a single point-to-point circuit. The term link can be used whether the point-to-point link is the entire circuit or only a single piece of a larger circuit or network.

The term *circuit segment* is used when the point-to-point circuit is a single piece of a larger multipoint circuit or network.

DATA LINK

Be careful not to confuse the term *link* with the term *data link*. Link refers to the physical circuit; data link refers to the physical circuit plus all of the equipment necessary for two devices connected to the circuit to communicate. A data link is usually not considered to have been established until the routines that establish data communications (called *hand-shaking* routines) have been completed and until all equipment is operating and ready for data to be transmitted.

NETWORKS

The term *network* is used to describe a connected system of circuits and their associated data communications control equipment. The term is sometimes applied to a single point-to-point circuit, but it usually refers to a more complex system consisting of a number of circuits. In other words, a network can be the data communications line used to connect one terminal to a computer; or a communications system a company installs within its own building to connect various word processors, CRT terminals, and computers (this is called a *local area network* or LAN and will be discussed in more depth in Chapter 10); or a system a company might use to connect terminals in its remote branches to the computer at corporate headquarters; or an airline's worldwide data communications system that connects thousands of terminals in countries all over the world to the airline's main computers, which in turn may be located in several different sites.

PATHS

A *path* is a particular route through a network. When two or more links are connected at one or more nodes (these are the connection points on a network), such as nodes A, B, and C in Figure 3-2, the multipoint circuit that results—that is, AB-BC—is often called a path or route. For convenience, a single link may also be considered a path or route.

CHANNELS

The term *channel* is usually used to indicate a one-way path for communications and does not refer to the physical medium itself. For example, a four-wire circuit consists of two pairs of wires, one for transmitting and one for receiving. Thus, each pair of wires supports one communications channel. A basic four-wire circuit, then, consists of two channels, a transmission channel and a reception channel.

Reverse Channels

When one of a pair of channels is designed to operate at a much lower speed than the other and is used for a lower volume of information (for example, for control information, which is the information that, among other things, indicates

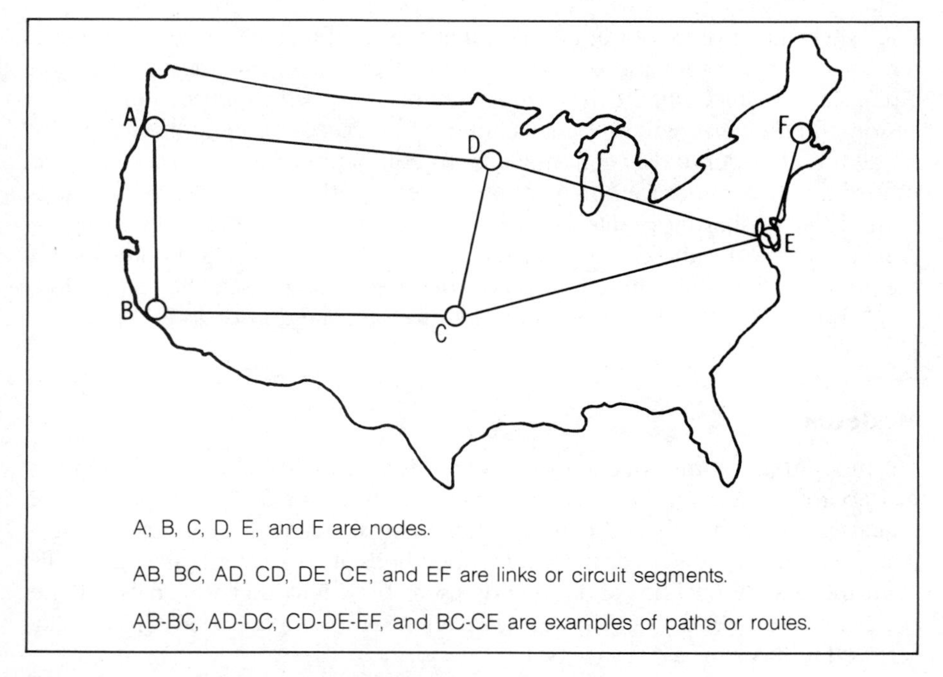

Figure 3-2. Nodes, Links, and Paths Through a Network

whether a block of data has been correctly received at the other end), that channel is called the *reverse channel.* The higher-speed channel is called the *main* or *primary channel* or the *forward channel.*

Muliple Channels on One Circuit

With most transmission media, the physical circuit can carry more than one simultaneous data conversation by the use of a technique known as *multiplexing.* Multiplexing will be discussed in detail in Chapter 8; for now you just need to realize that a technique exists for transmitting two or more separate data conversations on one circuit (for example, on one pair of wires). It is possible to multiplex many data channels on a single circuit, thereby improving the utilization of the physical circuit and usually lowering the data communications cost.

PHYSICALLY CONNECTING THE DATA TERMINAL EQUIPMENT TO THE TRANSMISSION CIRCUITS

Four specialized pieces of equipment that are used to connect a computer, a terminal, or any other piece of data terminal equipment to the transmission circuits are the *modem,* the *line controller,* the *front-end communications processor,* and the *hardware interface.* Of these four pieces of equipment, only the modem is considered data communications equipment. The line controller and the front-end communications processor are considered data terminal equipment, while the hardware interface can be part of either the DTE or DCE. Figure 3-3 shows how a remote computer terminal accessing a host computer over a telephone line may use these four pieces of equipment to make the connection.

Let us review the definition of modem we presented in Chapter 2.

Modems

The modem (the name is a contraction of *mo*dulator/*dem*odulator) is equipment that accepts data from the computer, terminal, or other device in the form of digital electrical signals and converts those digital signals into an analog form that can be transmitted over voice telephone lines; it also accepts analog signals from the telephone lines and converts them back into digital signals for the receiving computer device. The regulated telephone companies have historically referred to modems as "data sets."

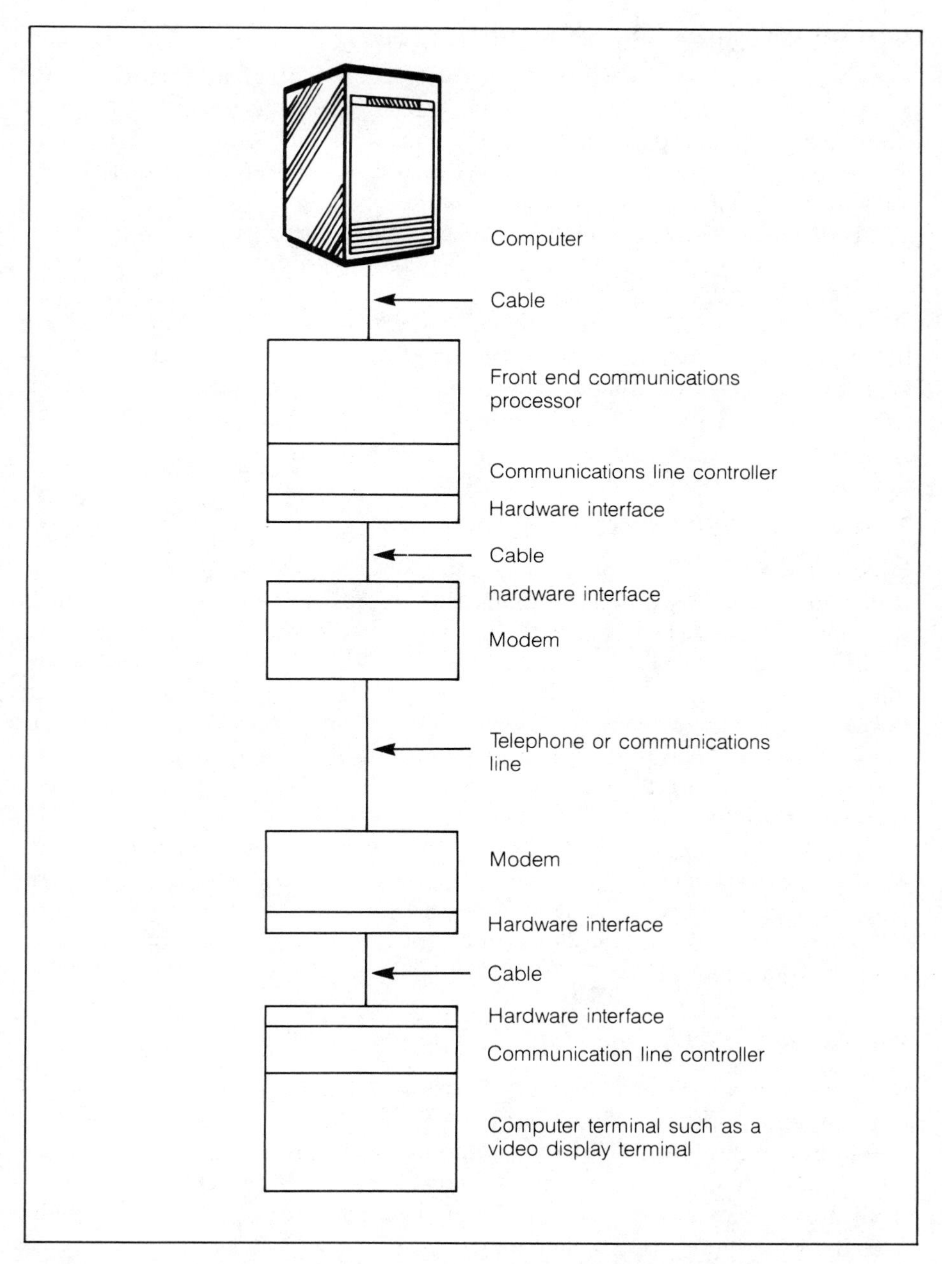

Figure 3-3. The Devices That Connect Data Terminal Equipment to a Transmission Circuit

Communications Line Controller

The *communications line controller* is a piece of computer hardware that, along with its associated software, accepts the data from the terminal keyboard or computer memory and performs the functions necessary to pass that data on to the modem. For example, it converts each character to the stream of bits needed for transmission and passes the appropriate control signals to the modem.

The communications line controller is usually integrated with other hardware; in some equipment, for example, the communications line controller's function is performed by the device's main control unit. You will usually hear about the communications line controller only in association with the computer, but in fact all peripheral devices connected to a modem (such as computer terminals) also have a communications line controller, although it may be much less sophisticated and perform fewer functions than the one associated with the computer.

Front-End Communications Processor

When a computer has many devices connected to it, the task of managing the data communications is quite complex and time consuming. This task includes deciding which device can talk with the computer at any point in time and which applications program each device is talking to; transferring messages that are intended for another device rather than for the host computer out to the other device; determining what to do if a transmission is not completed or has errors in it; and so on. The host computer can spend so much time managing the data communications that there is insufficient time available to handle the data processing load itself.

In these situations a *front-end communications processor* can be employed. This device is actually a specialized computer that is dedicated to handling the communications functions and thus frees the host computer for data processing. When a front-end communications processor is used, the line controller is normally integrated into it.

Hardware Interface

The *hardware interface* is the physical hardware that is used to connect electrically the communications line controller to the modem or modem substitute. The most common form of hardware interface for equipment used with micro- and minicomputer systems is a plug-and-cable system that adheres to the Electronics Industries Association's (EIA) standard RS-232C (the "C" refers to the latest revision of the RS-232 standard).

Most modems, computer terminals designed to be used with micro- or minicomputer systems, and micro- and minicomputers themselves have connectors sometimes called "ports" that conform to the RS-232C standard. A *port* is the communications access to the computer. The port consists of the physical hardware interface, such as the RS-232C connector, and the communications line controller.

Modems are thus connected to the computer or the computer terminal (actually they are connected to the communications line controller, which is inside the computer or terminal) by a cable with RS-232C connectors or "plugs" on each end.

In addition to the RS-232C, there are a number of other popular hardware interfaces. These include: the current-loop interface; CCITT's V.24 and V.35 (CCITT is an international organization, the Consultative Committee on International Telegraphy and Telephony, that recommends communications standards); EIA's RS-366A, RS-449, RS-422, and RS-423; the United States military's MIL-STD-188C (which is similar to the RS-232C but is designed to generate lower levels of electromagnetic radiation; IEEE's 488 parallel interface (the IEEE is the Institute of Electrical and Electronic Engineers); and the Centronics parallel interface.

Bit-Serial Interfaces Most data communications over the hardware interface is *bit serial*—that is, a series of bits is transmitted one after each other over a single channel.

Parallel Interfaces Another form of transmission that is especially popular between a computer and an attached printer is *parallel transmission.* In parallel transmission, an entire character (or group of bits) is transmitted at one time by sending each bit in the character or group over a separate wire or *pin.*

How the Hardware Interface Is Specified The definition of the hardware interface consists of at least three components:

1. The specification of the physical hardware—for example, the number, size, shape, and layout of the connecting pins
2. The specification of the electrical signal characteristics
3. The functional descriptions of the various circuits provided by the interface

You should note that all three areas are not always defined for a specific interface. For example, the name "RS-449" refers both to a set of specifications and to the hardware that is the interface itself. The RS-449 specifications do not specify voltage levels. With the RS-449, you will use one of two additional specifications, the RS-422 or the RS-423. Each defines the voltage levels to be used with the RS-449 interface (under different conditions), but neither the RS-422 nor the RS-423 represents a separate physical interface.

Popular Hardware Interfaces The following discussion will give you some familiarity with the more popular interfaces.

The *RS-232C* is the most widely used data communications interface in the United States. It has become *the* interface used with most modems designed for the voice telephone network and with terminals connected to mini- and microcomputer systems.

The RS-232C standard specifies that cables used to connect two devices can be no more than fifty feet in length, at the maximum data rate of 20K bps or 20,000 bits per second. It utilizes a twenty-five-pin connector. RS-232C specifies the functions that will be supported by the interface and is designed so that only one function is associated with each pin in the interface. Therefore, a large number of pins is required to support the many functions to be signaled over the RS-232C interface.

The *CCITT V.24* standard is the European or international equivalent of the RS-232C and is very similar in most respects.

The *20 mA* (milliampere) *current loop* is an interface that has been widely used in connecting teletypewriters and is often found on other types of terminal devices as well. Although once very popular, it is being phased out and replaced with other types of interfaces, particularly the RS-232C. You should note that there are also current-loop interfaces that utilize current levels other than 20 mA.

The EIA-specified *RS-449* interface is relatively new and is intended to replace the RS-232C. However, the RS-232C is so widely used that, even with all of the RS-232C's shortcomings, the RS-449 is having trouble gaining acceptance. The RS-449 specification calls for a thirty-seven-pin connector for the main channel and an optional nine-pin connector for a secondary or reverse channel. It is designed for data rates of up to 2 million bits per second (2M bps) and permits cable lengths of up to 4,000 feet.

The RS-449 specification does not cover voltage levels (the RS-232 does). Instead, two companion specifications, the RS-422A and the RS-423A, cover the specified voltage levels for data speeds in the 20K bps to 10M bps range and the 0 to 20K bps range, respectively.

The *CCITT V.35* interface uses a thirty-four-pin connector and is designed for higher-speed applications than the RS-232C or V.24 interfaces. The V.35 interface is used as part of CCITT's recommended X.25 packet-switching protocol for speeds higher than 9600 bps.

The RS-366A interface uses a twenty-five-pin connector for auto-dial applications and goes between the data terminal equipment and the automatic calling unit (ACU). (The ACU performs the functions of seizing a telephone line, dialing the desired number, and so on.) The international equivalent of the RS-366A is CCITT's V.25.

The *CCITT X.21* interface is designed for use between data terminal equipment and data communications equipment operating on public data networks (PDN). It uses a fifteen-pin connector. Unlike the RS-232C or the RS-449, the X.21 uses the same interface circuit for multiple functions. Four major circuits allow the specification of twenty-eight unique DTE/DCE states, such as "DTE waiting" or "DCE clear confirmation."

The *X.21-bis* is a second interface standard recommended by CCITT as an interim solution for use between DTE and DCE operating on public data networks where the DTE and DCE were designed for the RS-232C or CCITT V.24 interfaces. The X.21-bis interface looks like a V.24 interface to the DTE and DCE. The "bis" suffix is not an abbreviation or acronym; it merely denotes a second standard.

The *Centronics parallel* interface is designed by the Centronics Computer Data Corporation for use with their line of printers. It has become somewhat of a de facto standard parallel interface for connecting many manufacturers' printers to mini- and microcomputer systems. It utilizes a thirty-six-pin connector.

The *IEEE 488* is another parallel interface standard that is sometimes used in connecting printers to mini- and microcomputers. However, the IEEE 488 interface is most popular as an interface for the remote control of and data acquisition from test instruments such as voltmeters and oscilloscopes. The IEEE 488 interface uses a twenty-four-pin connector.

Because the IEEE 488 is a parallel interface, it will often be referred to as the *IEEE 488 bus*. It is also called the General-Purpose Interface Bus (GPIB) and the Hewlett-Packard Interface Bus (HPIB). Note that the GPIB and HPIB deviate slightly from the IEEE 488 standard. The IEEE 488 interface is available as an option on most mainframes, minicomputers, microcomputers, and personal computers.

BITS AND BAUDS

Here we look at some of the most confusing terms in data communications jargon—bits per second and baud rates. These terms have been misused or used with multiple meanings so much that these incorrect meanings have become common usage. I will explain the correct usage of both terms, but you should be tolerant of and expect confusing usage of them in data communications literature.

Bits

A *bit* is an abbreviation of "binary digit," which can be either a 0 or a 1. In digital communications, a bit is therefore the basic unit of information. *Bits per second* or *bps* are the number of bits or 0's and 1's that pass a point on a communications channel each second.

Start and Stop Bits

In asynchronous transmission (which will be discussed later in this chapter), signaling bits called *start and stop bits* are added to each character. Sometimes the bit-per-second rate refers only to the number of information bits transmitted per second—that is, it excludes the signaling or start and stop bits transmitted.

For example, if ten eight-bit characters, each with one additional start and stop bit, are transmitted each second, then the bit rate on the line is 100 bps (ten characters per second times ten total bits per character), but the "information bit rate" is only 80 bps (ten characters per second times eight bits of information per character).

Net Data Throughput

People sometimes also talk about the *net data throughput* (NDT) of a channel. NDT is usually expressed in either characters per second or bps. Net data throughput is the number of "usable" data characters or bits that are received per second and does not count characters that have to be retransmitted due to errors, characters used for control purposes, and so on.

However, in normal usage, the bit rate (in bits per second) refers to the total number of bits being transmitted per second.

Maximum Transmission Rate

With asynchronous transmission, characters are usually sent at less than the maximum rate the data link is designed for. But the bit speed of an asynchronous data link is usually specified at its maximum design speed. If someone says they are transmitting at 1200 bps, they mean they are sending 1200 bits down the line each second *when they are transmitting continuously*. However, due to the normally intermittent nature of asynchronous (and, to a lesser degree, of synchronous) transmission, over a period of 100 seconds they may in fact transmit far less than the 120,000-bit capacity of the data link.

Bauds

A *baud* is another measure of the speed of a data communications line. The term was named for the Frenchman Emile Baudot, a data communications pioneer, and refers to the number of signal events, or signal elements (that is, changes of specific conditions or characteristics that represent data) passing a point on the line, per second. If each signal event or element, such as a change from zero to ten volts, represents one bit, then the baud rate and the bit rate are the same. If one signal element can represent two bits (this is called a *dibit*), then the bit rate is two times the baud rate. For example, if 1,600 signal events occur each second but, through coding techniques, each signal element represents three bits (called a *tribit*), then the data signal is being sent at 1600 baud but at 4800 bps. This is the typical case with modern 4800 bps modems.

Keep in mind that people will often use baud where they should use bps or will use one of these two terms with a meaning somewhat different from the definitions provided here. So remember what the two terms mean, but expect to find them used frequently with somewhat different meanings.

TRANSMISSION CODES

If a series of bits are grouped together, say in groups of five or six bits, each of the many possible unique groupings of bits can be assigned to represent a character or symbol. These groupings of bits and their uniquely assigned symbols are called *transmission codes* or simply *codes*.

Control Characters

In addition to defining codes for all of the letters of the alphabet and the ten digits (the numbers 0 through 9), we need to define punctuation marks and special characters such as control characters. Control characters are groups of bits that have special meanings for a particular process such as data communications. Examples of common control characters are:

- SOH—start of header
- EOT—end of text
- BEL—used to ring a bell to get the operator's attention

Groups of Bits

The more bits in a group, the more unique symbols that can be represented directly by a group of bits. Therefore, a five-bit group (used in Baudot code, which is defined in the next subsection), can represent only thirty-two different symbols, but an eight-bit group, called a *byte,* can represent 256 unique symbols.

On the other hand, if fewer bits are used in a group, then more characters can be transmitted per second for any given bit speed. Thus, if a line is transmitting at 1200 bps, either 300 four-bit characters or 200 six-bit characters or 100 twelve-bit characters can be transmitted per second. Consequently, a compromise must be reached when a transmission code is chosen between the number of unique symbols that can be directly represented and the number of characters per second that can be transmitted at a given bit rate.

We will now look at a number of the more popular codes used in data communications.

Baudot Code

The *Baudot code* is the code used by many teleprinter or telegraphic systems throughout the world and is very similar to another widely used telegraphic code known as the International Telegraph Alphabet #2. Baudot code uses five bits per group (sometimes referred to as a five-level code), which provides for thirty-two different symbols. Since the alphabet uses twenty-six characters and the numbers zero to nine use ten characters, the five-bit code cannot be used

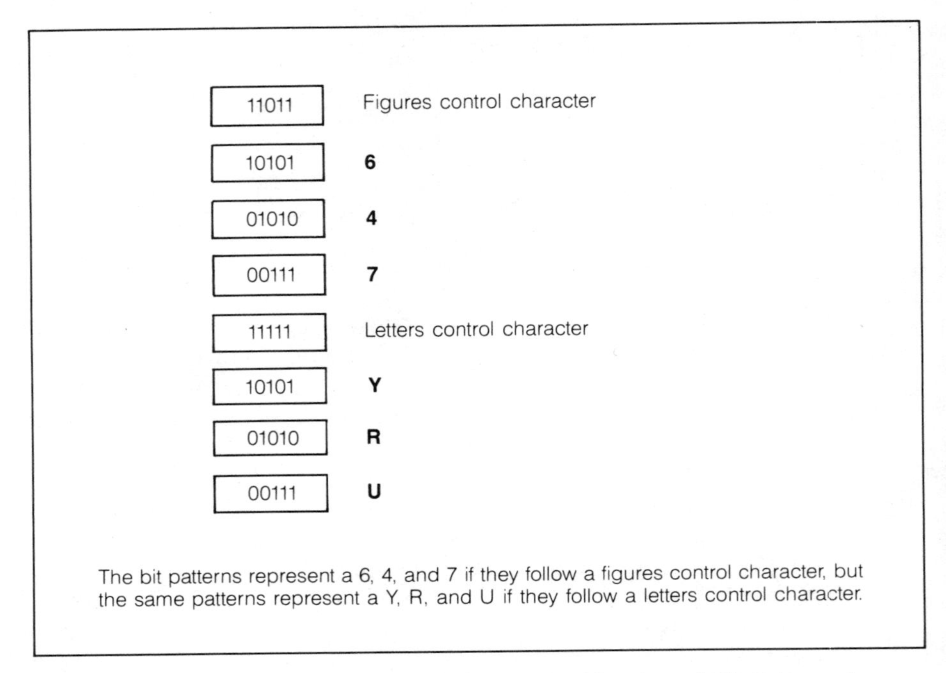

Figure 3-4. How Control Characters Change the Meaning of Bit Patterns in Baudot Code

to represent directly all letters and numbers, let alone any punctuation marks or special characters.

This problem is solved in Baudot code by using two of the thiry-two characters as control characters to indicate figures and letters. After the letters control character is sent, all the characters that follow are interpreted as letters until the figures control character is sent, and then all the characters that follow are interpreted as figures (see Figure 3-4). Thus, in Baudot code, a group of five bits will represent one of two unique symbols depending upon whether it is to be interpreted as a letter or a figure.

This type of code is called a sequential code, since the interpretation of a group of bits depends on its sequence—that is, whether it follows a letters or a figures control character.

Extended BCD

Extended Binary Coded Decimal, also called *extended BCD* or *EBCD* (and sometimes simply referred to as *BCD*) uses six bits to code a character. However, it also uses a parity bit and therefore takes a total of seven bits per character. (It is still referred to as a six-level code, as only six bits are used to code the characters themselves.)

Extended BCD was originally developed for IBM Selectric typewriters so they could transmit information over a communications line while simultaneously printing the same information or so that they could be used as terminals on timesharing systems. This code is also used with the IBM Displaywriter for data communications.

The term extended BCD is somewhat of a generic term; it has been applied to six-level codes that are basically similar but have some different codes for various characters. The Paper Tape Transmission Code (PTTC) and the Correspondence code are both EBCD codes.

ASCII

The American National Standard Code for Information Interchange, known as *ASCII,* is a seven-bit code that employs an eighth bit for parity. The seven bits allow the definition of 128 unique characters.

ASCII is used extensively with mini- and microcomputer systems and is the predominant code used with asynchronous transmission, since it defines a large number of punctuation and other special characters and provides a parity bit for each character. (Asynchronous transmission is a character-by-character transmission scheme that is used in most mini- and microcomputer systems to communicate with their terminals; it is discussed more thoroughly below.)

EBCDIC

Extended Binary Coded Decimal Interchange Code or *EBCDIC* is an eight-bit code that allows the specification of 256 characters. EBCDIC is used extensively in IBM mainframe computers and is often used with sophisticated protocols that check an entire message for errors and therefore have no need for a parity bit on each character. Consequently, EBCDIC does not utilize a parity bit.

Modified Code Sets

There are, of course, other code sets besides the four discussed here. Of the four, and in fact of all the data communications codes, Baudot, ASCII, and EBCDIC are the most popular in use today. But it is important to realize that, with transmission codes, ASCII is not always ASCII, EBCDIC is not always EBCDIC, and so forth. Some users have modified these codes or use a sequence of two or more symbols to have a special meaning. For example, some users of EBCDIC have eliminated certain of the special characters they don't intend to use or have redefined those characters and added a parity bit instead, thus obtaining an eight-bit code including a parity bit that is based on the EBCDIC coding scheme. Others have added definitions for some of the undefined bit groups. Still others have redefined certain bit groups to have different meanings on their systems.

Thus, when you are told a particular installation uses EBCDIC or ASCII, you must be sure they are not using a modified version of the official or standard code.

ASYNCHRONOUS AND SYNCHRONOUS TRANSMISSION TECHNIQUES

When data are transmitted digitally, each character is translated into a binary representation of that character that is based on the particular code being used. If ASCII is used, for example, then each character is translated into eight bits, a seven-bit code plus one parity bit. A sequence of characters such as words in a letter or a series of numbers will therefore be transmitted as a long series of bits or 0's and 1's.

Figure 3-5 shows the boundaries of each character by vertical lines in the top of the figure, but in the bottom of the figure you can see that all of the bits run together. The communications equipment at the receiving end needs to know where to separate the bits into characters, or where to "frame" the bits, so that they can be interpreted properly. The two transmission techniques used to accomplish this separation are *asynchronous* (async) and *synchronous* (sync) transmission.

Asynchronous Transmission

Character Framing Asynchronous transmission is a *character-framing* scheme; that is, the data are transmitted so that each character is framed with delimiters or markers that say, "This is the beginning or end of the character." The framing is done by the use of start and stop bits. In async transmission, a start bit is

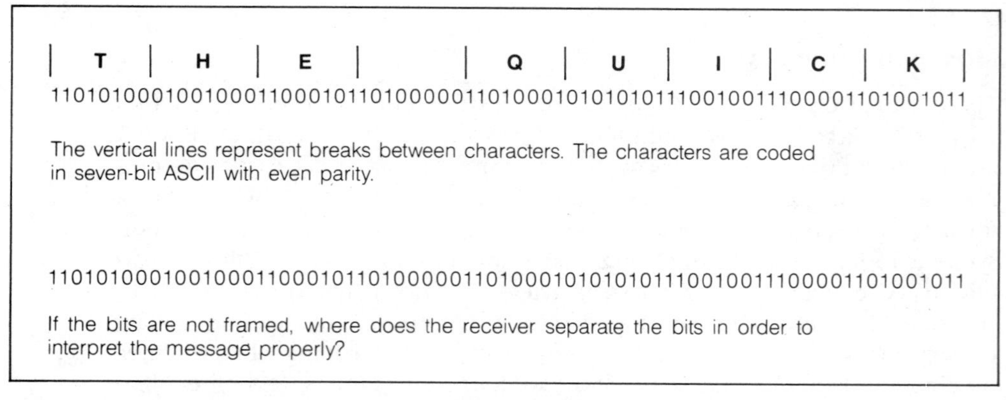

Figure 3-5. Why Sequences of Bits Need to Be Framed

transmitted at the beginning of every character and one or more stop bits is transmitted at the end of each character.

A start bit is a 0 or a *space*. A stop bit is a 1 or a *mark* (the terms mark and space come from telegraphy). When the receiving device sees a start bit, it then counts eight bits (if an eight-bit code like ASCII is being used) and looks for the stop bit or bits. It then waits for the next start bit, counts eight bits, and so forth.

When transmission starts, the receiving device (such as a modem) looks for the beginning of a start bit, followed by the correct number of data bits, followed by a stop bit. It may have to keep shifting over one bit at a time until it sees the correct sequence. This process of finding the correct starting point is made easier by the fact that there are usually some time intervals between successive characters in asynchronous transmission. There will normally be more than one "bit time" between the stop bit and the next start bit, which will help identify the start bit. This process is called *getting in sync* (even with async transmission).

Synchronization In asynchronous transmission, the start and stop bits provide both *bit synchronization* and *character synchronization*. For bit synchronization, the transition from a stop to a start bit tells the modem or terminal when to start looking for bits, or, in other words, allows it to "line up" the beginning of each bit. Remember that after the start bit, the following data bits (and the stop bit) all come one right after each other at exactly the required time. It is only entire characters that can be transmitted at any random time intervals, not the bits within each character.

For character synchronization, the start and stop bits clearly define the beginning and end of each character or, in other words, they tell the device where to "frame" the character.

Time Intervals Between Characters With asynchronous transmission, a series of characters can follow one right after another, or there can be a different time interval in between each character. Since each character can start being transmitted at any time interval after the last character, this mode of transmission is called asynchronous, meaning not in step with a clock or not clocked. The receiving device looks for the transition from the last stop bit to a start bit and doesn't care how long it has to wait. Therefore, this type of transmission is good for character-by-character transmission, which is characteristic of teletypewriters or interactive terminal applications.

One or More Stop Bits You should note that in asynchronous transmission, while there is only one start bit, there may be more than one stop bit depending on the particular scheme employed. Each scheme uses a specific defined sequence for start and stop codes. Many CRT terminals and mini- or microcomputer systems use one start bit and one stop bit. Baudot teletypewriters use one start bit and one-and-one-half stop bits, and Model 33 teletypewriters use one start bit and two stop bits. Note that the wait state in between the last stop bit and the next start bit is called a mark (the term *mark,* then, describes a stop bit, a 1, and the wait state).

High Overhead Asynchronous transmission is simple, requires no buffering (that is, it requires no intermediate storage of characters), and it allows characters to be transmitted, one character at a time, at any speed up to the maximum capacity of the data link. However, the overhead associated with this type of transmission is quite high. Because of the start and stop bits needed, it requires ten to eleven bits to transmit each eight-bit character. Synchronous transmission provides an alternative with lower overhead.

Synchronous Transmission

Synchronous transmission reduces this overhead by saving the characters (from the keyboard or computer) in a buffer to form a "message" and then transmits the message all at one time. Special control characters are used at the beginning and end of the message (which may be of varying length) much in the same way that start and stop bits are used at the beginning and end of each character in asynchronous transmission. These message delimiters frame the message and provide what is called "block" or "message synchronization."

Getting in Sync The sending and receiving devices need some way to get in character sync so that the receiving device can frame each character. Also, because the messages can be long, it is very important to control the timing of placing the bits on the line (bit synchronization). Thus, the effect of the transmission medium on the timing of the bits must be considered.

For bit synchronization, a series of alternating 0's and 1's is usually sent at the beginning of each transmission (this is sometimes called a "PAD" character).

For character synchronization, two or more special synchronizing or sync characters are transmitted at the beginning of each block of data and sometimes in the middle of long blocks. When a receiver sees at least two sync characters in a row, it knows the next group of bits (the length of the group will depend on the code being used) will either be another sync character or a properly framed data character.

Clocking Since the delineation of the message occurs at its beginning and at its end, synchronous transmission is referred to as a *message-* or *block-framed scheme.* Also, in contrast to asynchronous transmission, each character must be sent continuously within the block, and each bit must be sent at exactly the right time. A *clock* (a device that puts out a carefully timed digital pulse) is usually used in both the transmitting and receiving devices (such as modems) to ensure the correct timing. Most synchronous modems also have the option for the data terminal equipment itself to provide the clocking.

Some devices derive their clock from the 0 and 1 transitions in the data stream. This is called *self-clocking*. However, the more usual scheme is for the transmitting modem to send the clock signal to the transmitting terminal or computer to tell it when to send the next bit. If the transmitting device cannot synchronize itself to this "external clock" (the clock signal from the modem), then the modem has to be capable of buffering, or storing, the bits and transmitting each bit at the correct time.

Low Overhead By buffering the characters and transmitting them in reasonably long blocks without the need to insert start and stop bits on each character, synchronous transmission achieves a very low overhead when compared to asynchronous transmission. However, the associated communications equipment is more complex and therefore more expensive. Because of the nature of synchronous transmission, more complex message formats are necessary than those used with asynchronous transmission. See Figure 3-6 for an example of asynchronous and synchronous message formats. In this example, the SYN, STX, and EOT characters are eight bits each for thirty-two bits, which, even in this short message, is less than the forty-six start and stop bits required with asynchronous transmission. With a much longer message, only the same thirty-two bits for the SYN, STX, and EOT characters are required, although on some very long messages additional SYN characters can be used. Other formats will include reader information that requires additional characters. Nevertheless, in longer messages, synchronous transmission requires far fewer overhead characters (or bits) than asynchronous transmission.

\T/\H/\I/\S/ \/ \I/\S/ \/ \A/ \/ \S/\H/\O/\R/\T/ \/ \M/\E/\S/\S/\A/\G/\E/

Asynchronous message format

\= a start bit

/= a stop bit

These twenty-three characters (nineteen letters and four spaces) at eight bits each (seven-bit ASCII code plus one parity bit) require 184 bits. Twenty-three sets of start and stop bits require another forty-six bits

/SYN/SYN/STX/THIS IS A SHORT MESSAGE/EOT/

Synchronous message format

Figure 3-6. Asynchronous and Synchronous Message Formats

POINTS TO REMEMBER

This chapter has defined briefly most of the important terms that you will run across in reading data communications literature or in talking with vendors. We started by looking at basic data processing equipment (the computer and its peripherals); then we explained the difference between data terminal equipment (DTE) and data communications equipment (DCE). Next we covered the terms that describe how different elements of data terminal equipment are connected in order for data communications to take place: circuits, lines, links, data links, networks, paths, and channels. Then we looked at the four pieces of equipment necessary for connecting the data terminal equipment to the transmission circuits, and we also described many of the most common hardware interfaces.

The remainder of this chapter covered the specifics of transmission techniques: bit and baud rates, transmission codes, and asynchronous and synchronous transmission modes.

The next chapter defines another important term in data communications—protocol—and provides specific examples that show you more about how data communications works.

4

ESTABLISHING THE RULES OF DATA COMMUNICATIONS—PROTOCOLS

Protocols are the formal rules or conventions that govern the communications between two devices. A protocol is a set of rules that specifies in a very detailed manner what messages will look like (called the message format), how a data link will be established (called hand-shaking), how the messages will be transmitted (for instance, whether asynchronous or synchronous transmission is to be used and whether the transmission is to be half or full duplex), how the devices will determine when a message can be sent (whether someone else is sending a message at the same time or whether the receiving device is ready), how to determine if a message contains an error, what to do if there is an error, and so on. In a way, protocols are similar to the rules humans use for understanding languages.

COMPATIBILITY OF PROTOCOLS

Two devices cannot communicate if they do not understand each other's protocols. If devices have incompatible protocols, it is rather like one person talking in Russian and the other person only understanding English. But, just as people use translators so they can communicate with others who do not understand their language, protocol translators (equipment consisting of both hardware and software) exist that enable two computer devices to communicate even if they don't use a common protocol.

In other words, when you are purchasing data communications equipment, you not only have to be sure that the hardware interfaces are compatible between two pieces of equipment (as we explained in Chapter 3) but you must also be sure that the equipment is designed for the specific protocol you will be using.

DATA LINK CONTROL

Data link control or *DLC* is the implementation of a very important subset of all protocols. DLC is the combination of software and hardware in all the related equipment (such as the host computer, the communications processor, the computer terminal, and so on) that controls the communications line controller. This is the software and hardware that manages the transmission and receipt of the data over the communications line and looks for errors during this process.

DLC is responsible for such functions as connecting and terminating the data link, synchronizing the receiver with the transmitter, controlling when stations can send or receive data so that they do not interfere with each other, and detecting and correcting transmission errors whether by retransmission or other means.

There are other protocols besides data link control for data communications functions. Chapter 7 discusses some of these, including such examples as session control protocols. In the next section, however, we look at some specific examples of DLC.

SPECIFIC DATA LINK CONTROL PROTOCOLS

In the previous chapter we discussed the simple but high-overhead asynchronous protocol. Most of the advanced DLC protocols are synchronous.

Since one or more references to these protocols seem to crop up in many articles or discussions on data communications, you need to have at least a cursory understanding of them. In the following subsections, I will give you an introduction to binary synchronous communications (BSC) or bisync and then briefly look at synchronous data link contol (SDLC).

Binary Synchronous Communications

The *binary synchronous communications (bisync)* protocol was developed by IBM primarily to support communications between a host computer and its associated terminals. Bisync is also used for communications between computers, especially over dial-up lines between minicomputers and mainframes or other minis.

There are in fact a family of bisync protocols. So, just because a device operates bisync does not necessarily mean it can operate on the same multipoint line as other bisync devices.

How Bisync Works The bisync protocol works in the following manner in order to send a message from the host computer to a terminal:

- The computer sends a special "select" message to a particular terminal. The text of the message is one control character (ENQ), which means, "Can you accept data?"
- The terminal receives the ENQ and replies with a message whose text is the control character ACK, which means "acknowledged" or "yes."

- The computer then sends a message with the desired data as text.
- The terminal receives the data and sends the computer a message with the text "ACK," if the data have no detectable errors, or "NAK" if an error is found.
- If the response was NAK, the computer retransmits the data.
- If the response was ACK, the computer transmits a message with a new block of data as the text.

Since bisync is a select-and-hold protocol, the data link is held until transmission is completed. This means that the ENQ sequence is only used once for a transmission to a terminal even if many data block messages in a row will be sent to that terminal. If an accepted message was the last one to be sent to that terminal in that sequence (indicated by the end-of-text control character ETX), then the data link is terminated.

With the bisync protocol, the ACK control character that is sent actually alternates between ACK0 and ACK1 depending upon whether it is for an even- or odd-numbered block. This allows for the detection of duplicate or missing blocks.

Bisync Message Format A typical bisync message format is shown in Figure 4-1. The message starts with two SYN control characters, followed by a special control character SOH, meaning start of header. The STX control character means both start of text and, because it always follows the header, end of header. A variable-length text message is then transmitted that is ended by the control character ETB if more messages are to follow (this would mean to continue to

/SYN/SYN/SOH/...HEADER.../STX/...TEXT.../ETB or ETX/BCC/

SYN = synchronization character

SOH = start-of-header character

HEADER = header information of variable length

STX = start-of-text character (and end-of-header character)

TEXT = text information of variable length

ETB = end-of-text character—more messages to follow

ETX = end-of-text character—no more messages

BCC = block check character

Since bisync is a bit-oriented protocol, text and control characters are differentiated by special control characters that tell how to interpret the next character string. A line hit can cause one of these special control characters to be misinterpreted, and thus text can be mistaken for control characters or vice versa.

Figure 4-1. A Typical Bisynchronous Message Format

hold up the link—remember this is a select-and-hold protocol) or ETX if this is the last message to that terminal. If ETX is sent, then the data link is terminated. Finally a block check character (BCC) is transmitted.

Since bisync is a bit-oriented protocol, text and control characters are differentiated by special control characters that tell how to interpret the next character string. A line hit can cause one of these special control characters to be misinterpreted, and thus text can be mistaken for control characters or vice versa.

Bisync's Major Problem One of the big problems with bisync and other nonpositional data link control protocols is that the same positions in a message are used for text and for control characters. Thus, if a line error changes the bit pattern of a text character into the bit pattern of a control character, a serious problem can result. Character-oriented protocols like bisync are usually nonpositional.

Positionally Based Data Link Controls

The problem described above as well as others were resolved by a new generation of bit-oriented DLC protocols that are positionally based. These include high-level data link control (HDLC) and synchronous data link control (SDLC).

In positionally based protocols, specific positions in the message (separate control fields) are reserved for control characters. Thus, an ordinary line error is unlikely to cause a text character to be interpreted as a control character.

High-level data link control (HDLC) is the standard for advanced data link control protocols adopted by the International Standards Organization. Thus, it is the standard advanced protocol in Europe and most other foreign countries.

Synchronous data link control (SDLC), like bisync, is an IBM-developed protocol. It is a subset of HDLC. SDLC is intended to replace the bisync protocols eventually and offers significant improvements.

Some of SDLC's features are:

- SDLC can support half-duplex and full-duplex communications as well as a version of full duplex on a multipoint line where the host computer can simultaneously send to and receive messages from different devices. (Half and full duplex are defined in the next section.) Bisync, in contrast, supports only half duplex, which means that the communications line must be turned around each time the transmitting and receiving devices change roles.

- SDLC is bit oriented rather than character oriented and therefore independent of code structure, so devices with different transmission codes (such as the five-bit Baudot code and the eight-bit EBCDIC discussed in Chapter 3) can be mixed on the same line.

- Unlike bisync, SDLC does not use control characters but rather uses the position of the data—that is, where the bit patterns are in a sequence—to determine their meaning and to distinguish control data from text.

- SDLC can support point-to-point, multipoint, and loop architectures (these concepts will be discussed in Chapter 6). Bisync does not support loops.
- SDLC can transmit up to seven blocks of data without having to wait for a response as to whether the data were received correctly or not. This capability, along with the use of full duplex, means that SDLC can be much faster than bisync on long circuits. For example, on satellite circuits, SDLC is not as sensitive as bisync is to propagation delay—that is, the time it takes for the signal to travel from the earth to the satellite and back to the earth again.

Figure 4-2 shows an example of a typical SDLC message. The *flag* in this example is a unique character that signals the beginning or end of a frame or message. The flag at the beginning of the frame is immediately followed by an eight-bit identifier or *address* field and then an eight-bit control field. The text (if any) follows the control field and can be of variable length. The text is followed by a sixteen-bit cyclicol redundancy check character, which is followed by an eight-bit flag.

Since SDLC is a positionally based protocol, specific positions in the message are reserved for different types of information, such as address, control characters, and text. This means that a line hit cannot cause a text character to be misinterpreted as a control character.

HALF-DUPLEX AND FULL-DUPLEX TRANSMISSION MODES

As you can see from bisync and SDLC, various protocols support different transmission modes; for example, bisync supports half duplex and SDLC supports both half and full duplex. As we have found in other aspects of data communications, this terminology can be confusing. Let us look, then, at how these terms are applied to physical circuits, communications channels, modems, the devices that send or receive data (such as a terminal or computer), and to the type of protocol being used.

/FLAG	/ADDRESS	/CONTROL	/TEXT	/CRC	/FLAG/
8 bits	8 bits	8 bits		16 bits	8 bits

Figure 4-2. A Typical Synchronous Data Link Control Message Format

Half-Duplex and Full-Duplex Physical Circuits

The terms *half* and *full duplex* are sometimes used to describe the physical circuits themselves. When applied to dedicated physical circuits, half duplex means a two-wire circuit and full duplex a four-wire circuit. However, these meanings developed when medium-speed full-duplex modems required four-wire circuits. Now that modems are available that can operate in full duplex at up to 4800 bps on a two-wire circuit, this use of the terms half and full duplex should be avoided, but you may still find them used in this way in some data communications literature.

Half-Duplex and Full-Duplex Channels and Equipment

When applied to a communications channel, a modem, or a transmitting or receiving device, half duplex means that the channel, modem, or DTE will support communications in either direction but only one way at a time. Full duplex means that communications is supported in both directions simultaneously.

A word of warning is in order: some terminals have a switch marked "HDX/FDX" that changes the terminal from a local to a remote echo-back mode (see "Detecting Errors" later in this chapter). Now HDX and FDX are standard abbreviations for half duplex and full duplex, but this switch does not affect whether the terminal is communicating in one or two directions simultaneously. In fact, these terminals generally are half duplex—that is, they can only support data communications in one direction at a time.

Half-Duplex and Full-Duplex Protocols

When applied to protocols, full duplex means that the communications protocol is capable of handling communications in both directions simultaneously between the same two devices. Half duplex means that communication can only take place in one direction at a time.

The Confusion Surrounding These Terms

An example will illustrate the potential confusion. Figure 4-3 shows a simple network connecting a computer with a remote terminal. The circuit consists of a dedicated two-wire line; thus, the circuit itself is called, by convention, half duplex.

The user has terminals that can operate full duplex and has decided to use special 2400 bps two-wire full-duplex modems—that is, modems that are capable of supporting full-duplex transmission on two-wire circuits by creating two equal channels, each one at a different frequency.

Unbeknownst to the user, however, the communications protocol being used is half duplex—that is, it will only support transmission in one direction at a time. So we have a half-duplex protocol controlling data being sent through

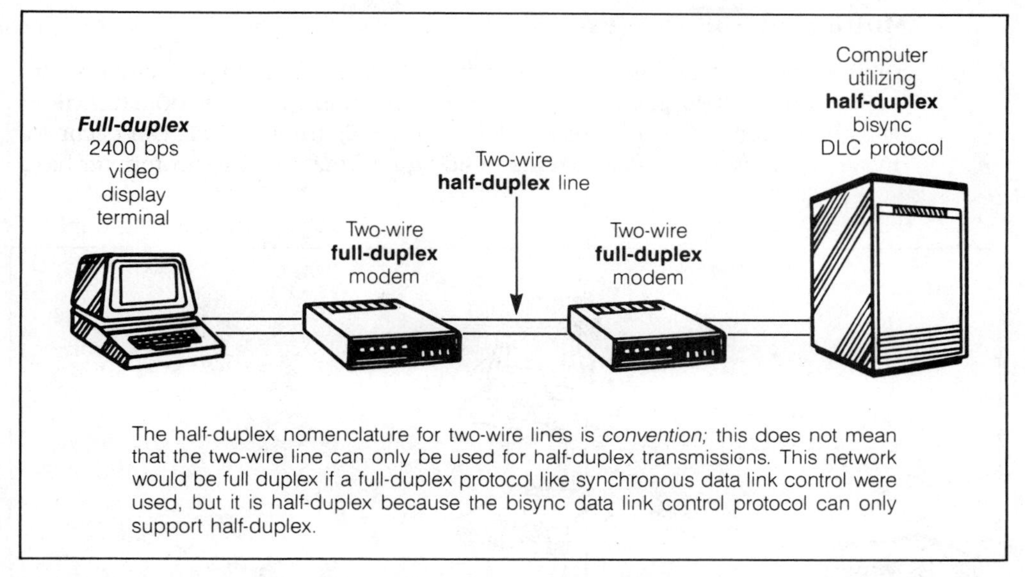

The half-duplex nomenclature for two-wire lines is *convention;* this does not mean that the two-wire line can only be used for half-duplex transmissions. This network would be full duplex if a full-duplex protocol like synchronous data link control were used, but it is half-duplex because the bisync data link control protocol can only support half-duplex.

Figure 4-3. An Example of the Half-Duplex/Full-Duplex Confusion

full-duplex modems operating over a half-duplex line between two full-duplex devices. That should be enough to confuse anyone! But the communications, in this case, would be half duplex, as the half-duplex protocol becomes the limiting factor.

Simplex

In addition to half duplex and full duplex, there is a transmission mode called *simplex*. Simplex means that communications can take place in one direction only. Commercial television transmission, radio transmission to the home, and transmission of data to a printer are good examples of simplex transmission.

International Definitions

Unfortunately, the International Telecommunications Union (ITU) uses different definitions of simplex and half duplex. ITU defines simplex as a circuit that can transmit a signal in either direction but not simultaneously (the same as the American definition of half duplex), and it defines half duplex as a circuit that is capable of full-duplex operation but that, because of the DTE or DCE attached to it, can only operate in one direction at a time.

So be extra careful when reading European data communications literature as either definition might be used.

Multipoint Full Duplex

There is another transmission mode that can be used on multipoint circuits. This is a version of full duplex where the simultaneous transmission in both directions takes place between two different devices (see Figure 4-4). This can occur for one of two reasons. First, the device sending information to the computer has a

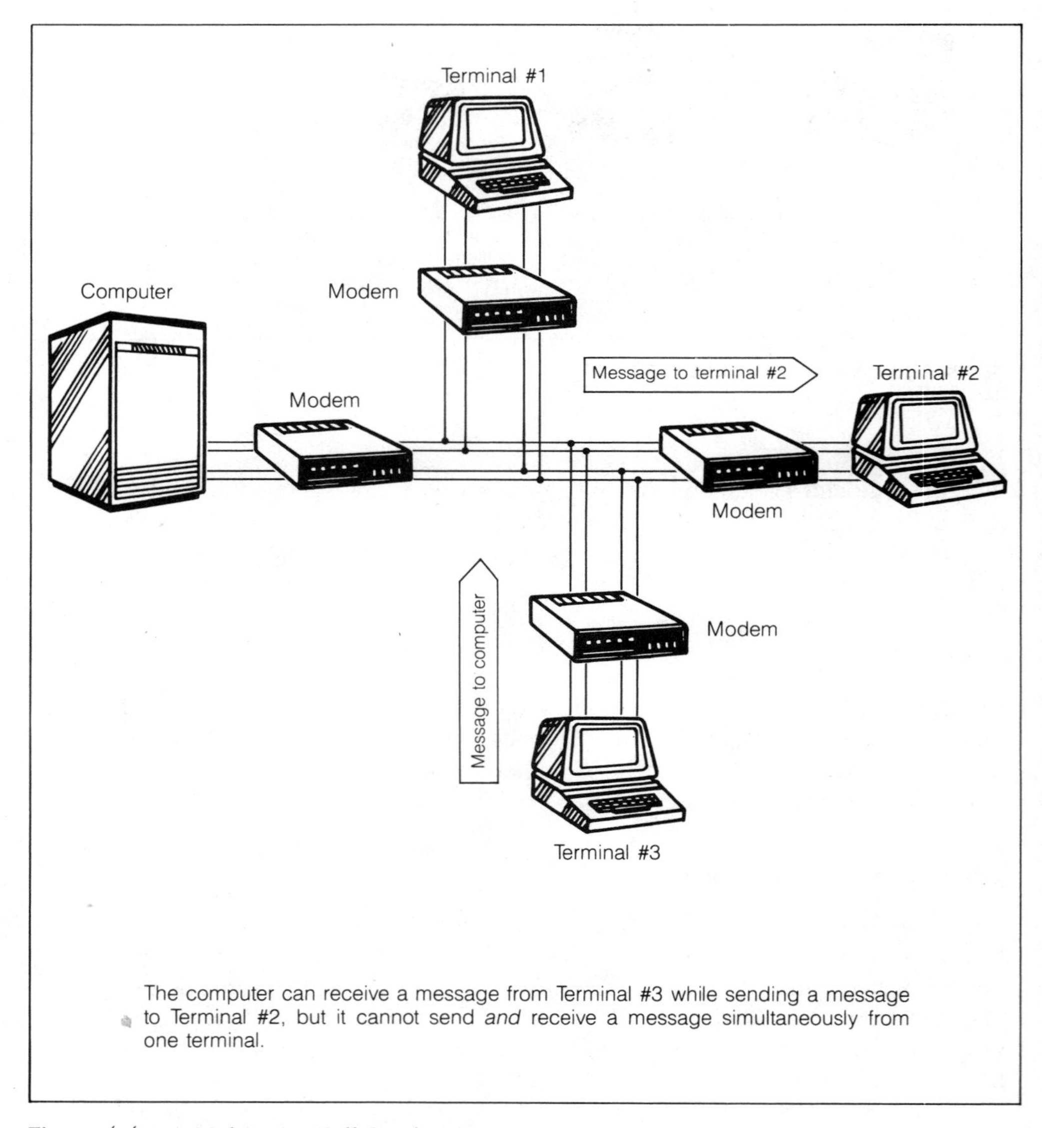

Figure 4-4. A Multipoint Full-Duplex System

long message, and, at the same time, the computer would like to send a series of short messages to several other devices on the line. If the computer can receive data from device 1 and simultaneously send other messages to devices 3, 6, and 8, for example, then the line utilization is improved, as is overall response time.

The second reason is that the devices attached to the line may not be capable of sending and receiving at the same time (for example, they may be too limited in buffer capacity or may not have the hardware or software to do it). In that case, if the computer, protocols, and network are capable of full-duplex transmission but the attached remote devices are half duplex, then overall network throughput can be improved by receiving from one device while transmitting to another at the same time. This has been called a restricted full-duplex protocol, a full/full-duplex protocol, a multi/multipoint protocol, and a full-duplex/multipoint protocol. You may see other names attached to it as well, and some will insist it is just a particular implementation of full duplex.

ERROR DETECTION AND CORRECTION

Protocols can provide for ignoring errors, flagging errors, or actually correcting them.

When transmitting data over more than a relatively short distance (typically 50 to 200 feet), the probability of errors being introduced during the transmission increases greatly. For instance, errors can be introduced on a data communications line by electromagnetic radiation, static electricity, or lightning strikes.

In digital communications, an error means that a bit that was transmitted as a 0 is received as a 1, or a 1 is received as a 0. Individual bits, whole characters, or blocks of data can also be "lost"—that is, never received.

Since the purpose of data communications is to transmit data from one place to another, techniques needed to be developed so that the data that were received were usable. The following subsections look at some of these techniques.

Ignoring Errors

Some protocols simply ignore the data communications errors. The theory here is that the operator would make more input errors than could be caused by the data communications process and that these data communications errors would be caught and corrected by the techniques developed to detect and correct the operator's input errors.

Detecting Errors

Most modern data communications systems, however, prefer to detect at least some errors, and they utilize some technique to correct those errors so that the applications programs are not burdened with handling transmission errors.

Echo Check A popular technique on slow-speed terminals (110 or 300 bps) is called *echo check, echo-plex,* or *echo back.* With this technique, each character is transmitted from the terminal to the computer, and the computer then transmits the character it receives back to the terminal where it is printed or displayed. The operator then verifies that the displayed character is the same as was typed originally, or corrects it.

Problems with the echo-check technique are that it can take a long time for every character to be received and then sent back and that, in its normal implementation, echo check requires an operator to look for the error and manually correct it. Obviously, the transmitting device could save the transmitted character and compare it with the echoed character in order to automate this function.

Echo checking also doubles the amount of traffic being sent; every character has to be sent twice (once in each direction). This increase in traffic is not usually a problem, however, in a manual input environment because the operator normally types much slower than the computer sends information. However, if a half-duplex line protocol is being used and the line must be turned around on each character, and if the line speed is low, then echo checking can interfere with an efficient throughput rate.

Automatic Repeat Request (ARQ) and Forward Error Correction (FEC) Another technique for detecting and correcting errors utilizes mathematical formulas to detect the errors. Once an error is detected, an automatic retransmission of the original data can be requested. This is called *automatic repeat request (ARQ)* or *automatic detection and retransmission.*

If the technique is sophisticated enough to determine which specific bit or bits are incorrect, then those bit errors can be corrected by the receiving data communications equipment. This is called *forward error correction (FEC).*

Parity The most widely used ARQ technique involves a system using "parity" to determine if an error has occurred. The parity systems add extra bits to each character (or group of bits), to the message, or to both.

The simplest form of parity is "one-dimensional parity" or *vertical redundancy check (VRC).* With VRC, one bit is added to each character or group of bits. The rules of VRC are:

- For *even parity,* an extra bit is added at the end of the bits representing the character so that the sum of all the 1's in the character including the parity bit is an even number.
- For *odd parity,* the sum of all the 1's must be odd.

Note that the VRC approach can only detect an odd number of errors in a group of bits. If two 1's get changed to two 0's, the parity will still be correct even though there are two errors (see Figure 4-5). With the dibit modulation schemes used in 2400 bps modems, two bit errors have become more frequent

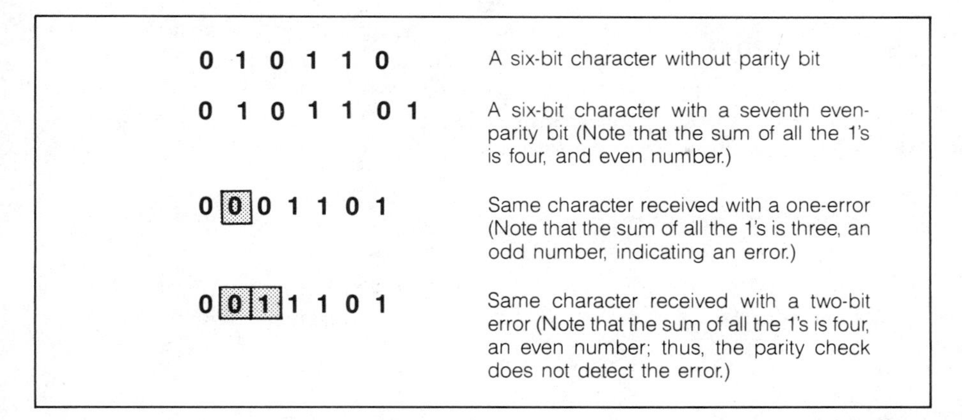

Figure 4-5. When Parity Checking Doesn't Work

than they were with 300 bps modems since two bits are being carried by the same signal element.

A second parity check can be added called a *longitudinal redundancy check (LRC)* or *block check character (BCC)*. This method is also known as a two-dimensional parity check. In addition to checking for parity on each character (or group of bits), parity is checked on each message or block by adding additional bits to the end of the message.

Cyclic Redundancy Check (CRC) In spite of the increased probability of detecting an error with both VRC and LRC, there are many errors that even two-dimensional parity will not detect. For this reason, a much more sophisticated error-detection technique called a *cyclic redundancy check (CRC)* was developed. The CRC approach uses a complicated mathematical formula and a specific polynomial to generate a very accurate check on the data being received.

Constant Ratio Code Another error-detection technique is called a *constant ratio code* or *X-out-of-Y* code. For example, IBM has used a 4-out-of-8 code, and a 3-out-of-7 code is used in radio telegraphy. This is another mathematical approach, but it is not as widely used as parity checking. In fact, this approach is often implemented in conjunction with LRC parity checking.

What Happens When an Error Is Detected

In the automatic repeat request system (ARQ), if no error is detected by one of these error-checking techniques, then the receiving device usually sends an ACK or acknowledgment to the sending device to indicate that the message was received and is error free. To save time and characters, some protocols do not send back a response if no error is detected.

If, on the other hand, an error is detected, a negative acknowledgment (NAK) is returned to the sending device. When the sending device receives a NAK, it knows that it must retransmit.

There are two widely used approaches to ARQ. One is known as stop and wait, while the other is called continuous ARQ.

Stop-and-Wait ARQ With *stop-and-wait ARQ,* the receiving device transmits one character or message at a time, then stops and waits for an ACK or an NAK before transmitting the next data. This is obviously a slow and inefficient use of the transmission facilities, especially if there is long turnaround time for the ACK, as with a satellite circuit.

Continuous ARQ This is a much more efficient technique but also much more complicated. With *continuous ARQ,* the sending device sends data continuously, without waiting for an ACK. If the receiving device detects an error, it sends a message back to the sending device. The sending device can then either resend the message in error (called "selective retransmission") or can resend the message in error plus all messages that were sent after it (called "go-back-N retransmission"). The go-back-N technique seems like an unnecessary transmission of extra data, but it eliminates the need for the receiving device to reorder the messages (see Figure 4-6).

Forward Error Correction (FEC) *FEC* utilizes extra bits in each character (or group of bits) not only to detect an error but also to detect precisely which bit is in error. Once the receiving device knows which bit is incorrect, it is easy to correct it since a bit can be only a 1 or a 0. FEC requires more redundant bits and more complicated algorithms than ARQ approaches and thus is more expensive to implement.

FLOW CONTROL PROTOCOLS

Sometimes a terminal or other piece of data terminal equipment will be transmitting data faster than the data link or other part of the network can handle the data. In this situation it is necessary to instruct the DTE to stop sending data temporarily. When the bottleneck has cleared up, the DTE has to be instructed to start sending again. This is called *flow control.* A simple example of flow control is when a printer runs out of paper. In this instance, the printer must be able to tell the computer to stop sending data until more paper has been put in the printer.

The next two subsections look at the major methods of flow control used on individual data links.

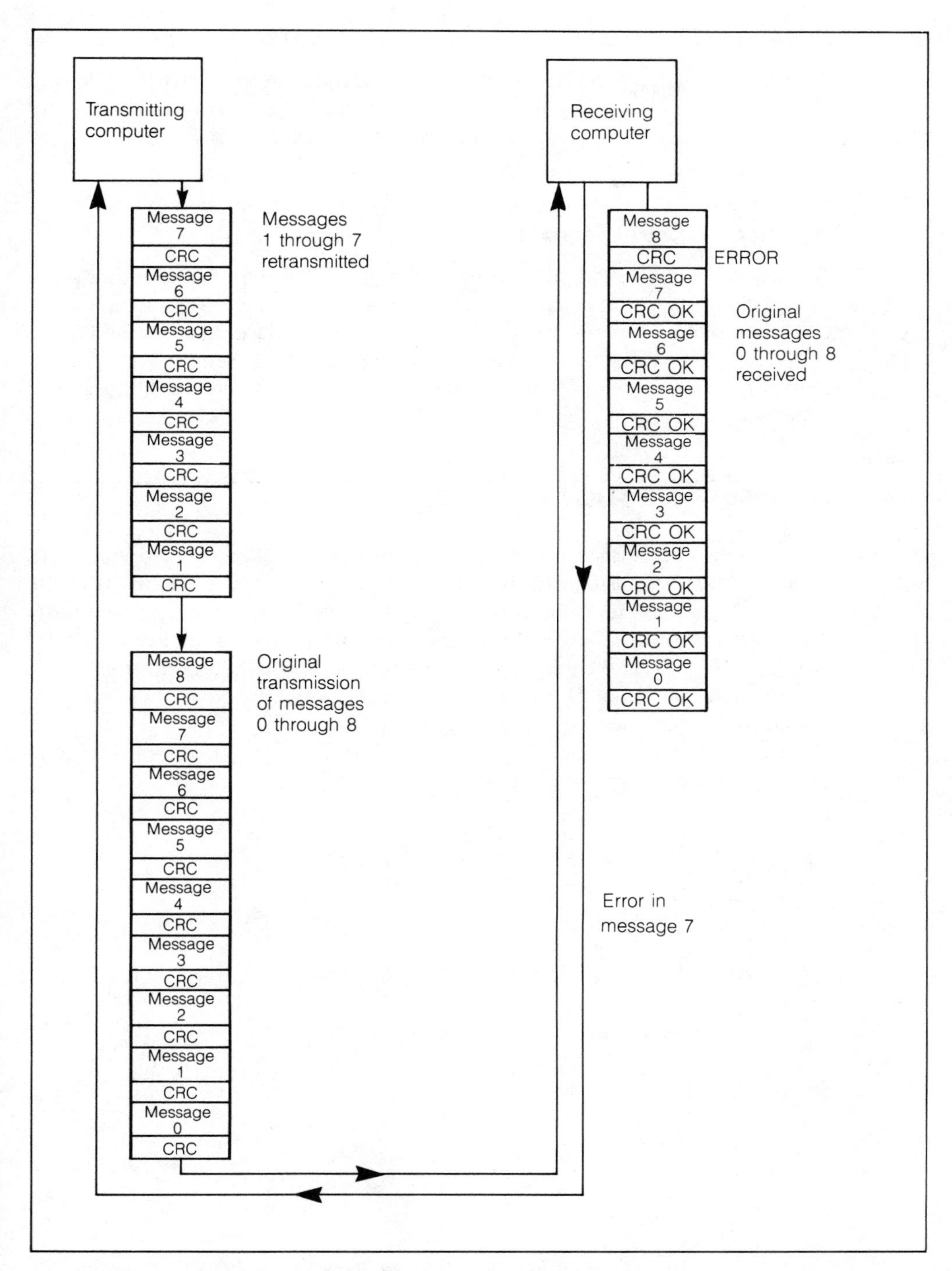

Figure 4-6. The Go-Back-N Technique of the Automatic Repeat Request Protocol

XON-XOFF

In the *XON-XOFF* method, the controlling device sends a special character (known as the XOFF character) to the DTE, which means "turn off your transmission." When data flow can start again, the special character used for XON is sent, meaning "turn on your transmission."

Interface Control Signals

The second major method used is to send a signal to the modem or other piece of data communications equipment that is connected to the DTE, instructing the DCE to send the correct signal over the hardware interface that tells the DTE to stop or start transmission. For example, raising the clear-to-send (CTS) signal on the RS-232C interface instructs the DTE that it is okay to start transmitting. The effect is the same as if the DTE had received an XON control character.

POINTS TO REMEMBER

Protocols are the language or rules computers use to talk to one another or to their peripherals. This chapter has given you a glimpse of the options available to you in data communications protocols. We have looked at examples of data link control protocols, error-detection/correction protocols, and flow control protocols. We have also explained half-duplex and full-duplex transmission modes and warned you about checking for protocol compatibility on the data communications equipment that interests you.

The next chapter looks at your options for the central piece of data communications equipment—the modem.

5

MODEMS

The *modem* is probably the single most important piece of data communications equipment. Virtually every data communications application requires a modem or similar signal-converting device (the only exception is when a device such as a terminal is directly connected to its computer). Because modems are an indispensable part of data communications over telephone lines or similar circuits and because modems come in such a wide variety of "shapes, sizes, colors, and flavors," this chapter examines modems and related devices at length. Here we give you the information you need in order both to purchase a modem and to ensure that it works for you once you have one.

THE MODEM AS A BLACK BOX

In Chapter 3 we defined a modem as follows:

The modem (the name is a contraction of *mo*dulator/*dem*odulator) is data communications equipment (DCE) that accepts data from the data terminal equipment (the computer, terminal, or other device) in the form of digital electrical signals, converts those digital signals to analog ones that can be transmitted over voice telephone lines, and reconverts the analog signals into digital ones at the receiving end.

In the computer field, and in the electronics field in general, we talk about "black boxes." Black boxes are devices that typically have an input and an output (or multiple inputs and/or outputs) and perform certain functions on the inputs before outputing the results. Black boxes are not usually black in color (although some are), but the term conjures up the image of an opaque box that you cannot see into and thus cannot figure out how it works. In other words, we are concerned only about what the changes are that the black box makes to the input signal but not *how* it makes those changes.

For the purposes of this book, I will treat modems and some related devices as black boxes. I will discuss what those devices do but will not discuss how they do it.

If you want to transmit data over more than a relatively short distance, such as fifty to two hundred feet, you have several options as to what kind of black box to use. The following sections describe some of those options.

DIRECT CURRENT LOOP

If you choose to use a direct current loop, no black box is required. You connect the wires directly to the current-loop interface on your data terminal equipment, and then you can run those wires several thousand feet. The current loop has been used primarily on teleprinters, teletypewriters, and similar data terminal equipment. It is normally implemented at 20 mA (that is, 20 milliamperes DC) and has also been widely implemented at 62.5 mA (usually referred to as 60 mA).

The use of the current-loop interface continues to decline. You will probably not consider it for most computer-related data communications applications, but many teleprinters and CRT terminals still use it. You may want to consider it for applications where a terminal must be located 200 to a few thousand feet from the computer if your computer has a current-loop interface.

LINE DRIVERS

Another approach is to use a black box known as a *line driver* (the telephone companies usually call these *local data sets*). A line driver is a black box that takes the digital electrical signals from the DTE and amplifies and possibly reshapes the pulses so that they can be transmitted farther than would be possible with RS-232C standards. Line drivers usually can handle data speeds of up to 9600 bps or 19.2K bps, and they can transmit over distances of up to several thousand feet or even longer.

Line drivers are inexpensive compared to conventional modems, but they do require DC continuity on a pair of wires. "DC continuity" means that a circuit appears to be a continuous circuit (rather than an open circuit) to a direct current. This usually means a continuous metallic circuit—that is, a circuit made of wire that is not interrupted by devices such as amplifiers, transformers, or capacitors. Many circuits in the voice telephone network look continuous to a voice signal but look open to a direct current because of the insertion of devices such as amplifiers or coils.

Because line drivers require DC continuity, they are usually used on circuits within a customer's building or group of buildings. In some cases, however, a pair of wires with the required DC continuity can be obtained from your local telephone company. Private-line metallic circuits (that is, circuits that have DC continuity) obtained from the telephone company are often called 43401 circuits, after the AT&T publication that describes the specifications for such circuits.

Line drivers can usually be used in a series with two or more at different points on the line so that longer distances can be covered.

MODEM ELIMINATORS

Some line drivers can emulate the control signals of a modem. These control signals are necessary in order to provide data link control for some terminals and computers. When the line driver is capable of producing these control signals, it is also said to be a *modem eliminator*.

Modem eliminators are also available as stand-alone black boxes that do not include the line driver. In this case, they provide the necessary DLC signals for the RS-232C interface between a computer and a terminal.

SHORT-HAUL MODEMS

The term *short-haul modem (SHM)* can be confusing since it is used inconsistently. Sometimes it is used as a synonym for line drivers. Sometimes it is used as a synonym for limited-distance modem (LDM), which is discussed in the following section.

Sometimes SHM refers to a black box that, like a line driver, operates only over a pair of wires with DC continuity and conditions the digital pulses so they can travel distances of up to ten or twenty miles. Since the pair of wires has a much larger bandwidth than a 4000 Hz voice telephone channel, this type of SHM does not have to modulate a carrier (see the discussion of modulating a pulse or voltage level in Chapter 2) in order to transmit the data but just applies the conditioned pulses to the metallic circuit. For this reason, these devices are also called "baseband" modems. (You should note that these devices are usually current-loop devices, although normally not the standard 20 mA type; thus, they are sometimes called "current drivers.")

Finally, sometimes SHM refers to a modem that utilizes a form of baseband modulation such as polar modulation (which reverses the direction of the current flow). By modulating the signal, these short-haul modems do not require DC continuity and can use circuits that are AC or inductively coupled (that is, connected by a coil or transformer-like device).

Whatever the exact use of the term SHM, this black box is capable of transmitting data signals over shorter distances than a conventional modem but also at a lower cost. SHM's can usually work over distances of from one to twenty miles and may or may not need DC continuity over a pair of wires depending upon the way the specific black box works. You should consult the manufacturer or distributor or read the technical literature to be sure about an SHM's suitability for your application.

LIMITED-DISTANCE MODEMS

Limited-distance modems (LDM) are usually conventional long-haul modems without some of the latter's electronic functions such as auto-equalization. LDM's typically work in the one-mile to 200-mile range depending upon conditions and the particular modem design.

LDM's are less expensive than conventional modems primarily because they use manual rather than automatic equalization. Since the conditions on even a dedicated circuit can change (water can get in the cable, for example, or sunspot activity can cause changes), you may have to readjust the equalization of an LDM to keep the circuit working.

As we mentioned in the preceding section, the term LDM is often used to describe baseband-modulated SHM's, so be sure you understand the limitations of the modem you are considering before you make a purchase.

CONVENTIONAL OR LONG-HAUL MODEMS

These are the modems you will usually use on voice-grade telephone circuits. The four-wire voice-grade private-line or leased data circuits obtained from the telephone company are often called *3002 four-wire circuits* or sometimes *41004 circuits* after the AT&T publication number that describes the specifications for them. As we've said before, the telephone companies who supply their own modems usually refer to their conventional modems as "data sets."

Modem Speeds

Modems normally come in the following line speeds: 110, 300, 1200, 1800, 2400, 4800, and 9600 bps (some modems are also available at 3600, 7200, 14,400 and 16,000 bps). As a rule of thumb, the higher the line speed, the more expensive the modem. This is because the encoding techniques are more complicated and because the electronic circuitry to compensate for line problems is also more complicated at higher speeds.

Let us look more closely at low-speed, medium-speed, and high-speed conventional modems.

Low-Speed Modems Low-speed (1800 bps or less) modems are usually asynchronous and may operate either over dedicated or dial-up lines or over both. Modems that operate over dial-up lines are normally half duplex, but special modems are available that operate full duplex over dial-up lines.

Low-speed modems usually use a technique known as frequency modulation or frequency-shift keying (FSK) to encode the digital signal. Higher-speed modems use phase-shift keying (PSK) or differential phase-shift keying (DPSK) as well as other techniques. Some modems employ secondary or reverse channels that operate at very low speeds (0 to 5 bps), which utilize amplitude modulation (AM).

It is not necessary for you to understand these different techniques, but you should realize that different modems use different modulation or encoding techniques and that some modems look for shifts in frequency and some modems look for shifts in phase. This will be important when analyzing transmission errors, as phase errors are, of course, more important to modems that use PSK or DPSK than to those that use FSK.

Medium-Speed Modems Medium-speed (2000 bps to 4800 bps) modems are usually synchronous devices, and, although modems in this speed range can be designed to operate full duplex over dial-up lines, most can only operate half duplex over dial-up lines and must have four-wire dedicated lines or two dial-up lines in order to operate full duplex.

Racal-Vadic was the first to introduce a 2400 bps full-duplex modem designed to operate over dial-up voice circuits (their model V24400). Other companies, such as Codex with the Codex 224 modem and Concord Data Systems with the CDS-224, followed shortly with their 2400 bps full-duplex modem. Then, just as the data communications community was getting used to being able to send 2400 bps full duplex over two wires, Anderson Jacobson introduced their model AJ4048 full-duplex 4800 bps modem designed for use over two-wire dial-up lines.

Medium- and high-speed modems usually use a phase modulation technique to encode the digital signal. Thus they are more sensitive to phase errors during transmission than are slower modems using other techniques.

High-Speed Modems High-speed modems are generally considered those that operate at 7600 and 9600 bps. However, Codex has introduced a modem that operates at 14,400 bps and General DataCom (GDC) has introduced one that operates at 16,000 bps.

High-speed modems usually operate over four-wire dedicated circuits that permit a full-duplex mode of operation. But Gandalf has a modem called "super-modem II" that operates full duplex at 9600 bps over two dial-up lines. It also operates over one four-wire unconditioned voice-grade dedicated line.

Modems that operate at 9600 bps can theoretically transmit a continuous stream of data twice as fast as 4800 bps modems. However, before you invest in 9600 bps modems you should try them out. Most systems do not transmit data continuously for long periods of time, and thus your throughput will be doubled only at peak times or on short bursts of data. High-speed modems are also usually more sensitive to transmission errors than slower modems. Your net data throughput, because of the extra time spent on error recovery, may not be much higher with the high-speed modem than with a 4800 bps modem if you have a noisy circuit.

ACOUSTICAL COUPLERS

Acoustical couplers are black boxes that are actually conventional voice-grade modems, but they do not connect to the telephone line by means of a hard-wired interface. Instead, they connect to the telephone line by placing the telephone handset into two rubber cups (one for the receiver and one for the transmitter) that "couple" the black box to the handset acoustically (see Figure 5-1).

Acoustical couplers are available only for low-speed transmission, and they come in speeds of 110, 300, and 1200 bps. They are particularly advantageous for terminals that are going to be moved from office to office. For example, if your business uses one or two portable terminals that connect to a timesharing system, are shared by five or ten people in different offices, and are sometimes taken home at night, you might want to use acoustical couplers.

When you use an acoustical coupler for long periods of time on the same telephone set, the carbon granules in the telephone handset transmitter may tend to settle, degrading transmission. This most often occurs with acoustical couplers that place the telephone handset on top, in the same position as if it were in its cradle on the telephone. Sometimes hitting the handset against a solid object like your desk will loosen the granules up, or you can replace the transmitter with a better-quality one.

WIDE-BAND MODEMS

Wide-band modems are available at speeds greater than high-speed modems, such as 19.2K bps, 56K bps, and 230.4K bps. These wide-band modems (for example, AT&T's 303 series) will not operate over voice-grade circuits but require a special and more expensive circuit.

Recently, modems that, by means of complex encoding techniques, operate over voice-grade circuits at speeds faster than 9600 bps have been introduced, such as the ones by Codex and Gandalf mentioned earlier. However, these are still referred to as conventional, rather than wide-band modems, since they operate over voice-grade lines and thus do not require special wide-band circuits.

DIGITAL SERVICE UNITS

There is a type of black box that is merely a signal converter and not a modem but is sometimes referred to as a *digital modem.* It is also known as a *digital service unit* and called by AT&T a *data service unit (DSU).* It is used to connect computers or terminals with an RS-232C or similar type of hardware interface

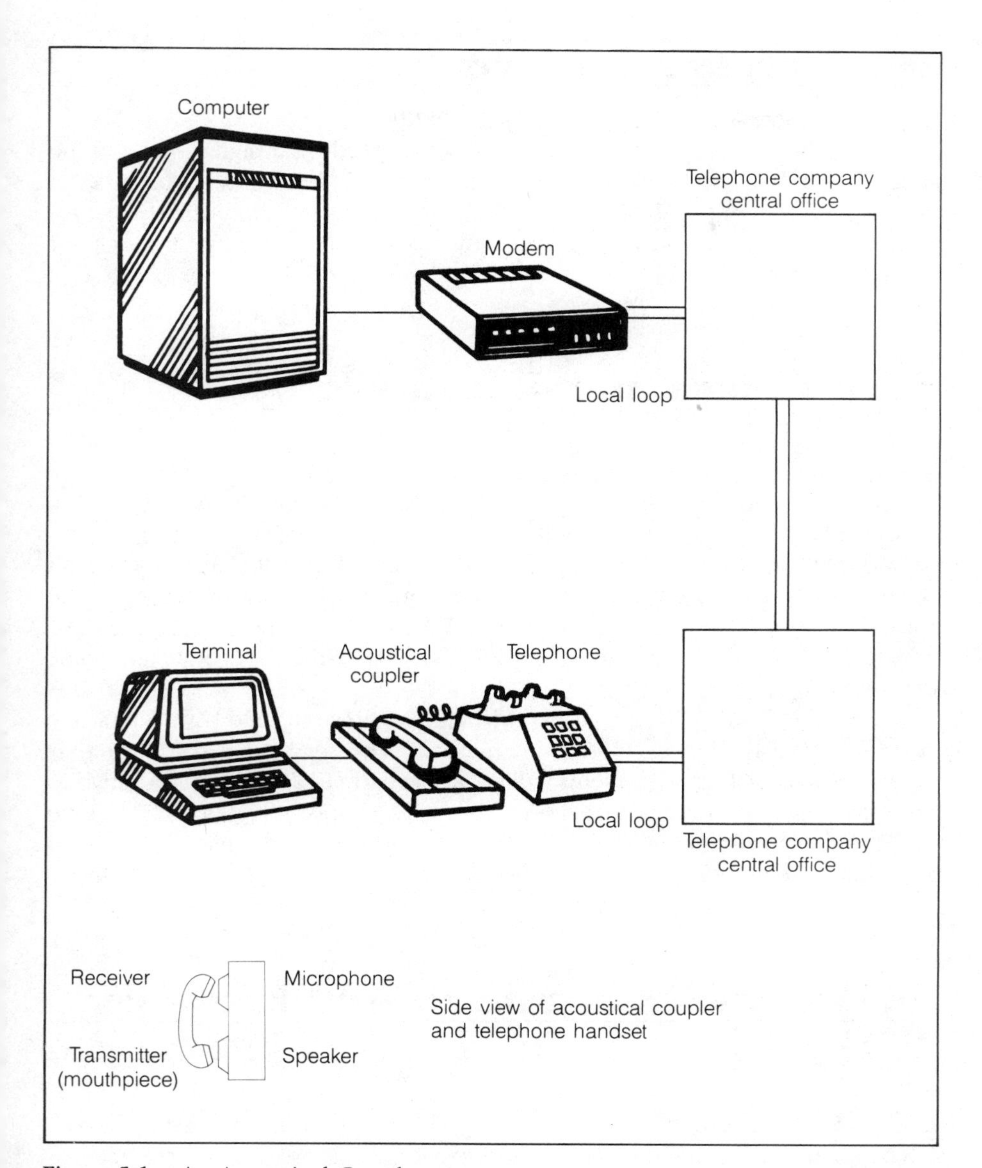

Figure 5-1. An Acoustical Coupler

to a digital transmission line such as AT&T's Dataphone Digital Service (DDS), a digital private-line service. Since the RS-232C signal is unipolar (the current flows in just one direction), a DSU is required to convert it to the bipolar signal (where the current flows in both directions).

CHANNEL SERVICE UNIT

If the user has data terminal equipment that can produce a digital signal in the specified bipolar format, then a DSU is not required, but a black box called a *channel service unit (CSU)* is required. The channel service unit provides protection for the network and provides the user with the network clock for synchronization.

DATA ACCESS ARRANGEMENTS

You may come across the phrase *data access arrangements* or DAA's in conjunction with modems. Prior to 1976 when the FCC instituted its registration program, all modems not supplied by the telephone company had to be connected to the telephone network by means of a black box called a DAA. The DAA was supposed to protect the telephone network from the harm that might be caused by the customer-provided modem. DAA's come in three types: the CDT, which is a manual device, and the CBS and CBT, which are both automatic devices.

As of June 1, 1977, all new modems (including those supplied by a telephone company) had to be registered with the FCC. The registration requirements include ensuring that the modems will not cause harm to the telephone network. Thus, a registered modem does not require a DAA. Since many of the pre–June 1977 modems will gradually work their way out of service, the use of DAA's will become less common.

CONNECTING MODEMS TO THE TELEPHONE NETWORK

Under FCC requirements, a modem is registered based on its method of connecting to the telephone network. Let us look at the four methods for connection, which correspond to the registration categories.

Acoustical Connection

As we discussed earlier, modems can be connected acoustically rather than hard wired to the network. In this case, they are usually called acoustical couplers and are connected by placing the telephone handset in the cups of the coupler.

Permissive Connection

Modems may be registered as "permissive" devices; this means that the amplitude of the signal presented to the telephone line must not exceed −9 dBm (decibels relative to one milliwatt). Permissive modems are normally connected with the same type of modular jack used with a regular telephone. This is called an RJ11C/W voice jack. If you also want to install a separate telephone on the same line so that you can alternate between voice and data use, then an RJ16X jack is used for the modem and an RJ36X jack is used for the telephone set.

Programmable Connection

If the modem is registered as a "programmable" device, it will typically be connected to the telephone network by a special programmable data jack known as an RJ45S. This jack can be programmed by the telephone company (they insert a resistor of the proper value) to set the amplitude of the signal presented by the modem from 0 to −12 dBm depending on the distance to the telephone company's central office. This allows the modem to transmit at its maximum allowable level without exceeding FCC-permitted levels on the telephone line.

When multiple modems are being used, a programmable connection known as the RJ27X is available that can connect up to eight modems each to their individual data lines.

Fixed-Loss Loop

The fourth way a modem can be registered is as a "fixed-loss loop" (FLL) device. FLL modems are limited to an output of −4 dBm. They use a connector known as an RJ41S data jack. The RJ41S with its switch in the FLL position will insert an 8 dBm attenuation pad in the circuit (this data jack is also known as the "universal data jack" as it will connect either FLL or programmable devices).

Unfortunately, the FLL type of connection tends to increase the circuit's susceptibility to a type of transmission impairment called "impulse noise." Consequently, I recommend that you check to see if a modem requires an FLL connection before you purchase it. If it does, consider a different modem if possible.

ERRORS DURING TRANSMISSION

A whole host of problems can occur during transmission of the data signal that can cause data to be lost entirely or a bit transmitted as a 0 to be received as a 1 (or vice versa). Some modems are designed to compensate for a number of these possible transmission problems. Or, for an additional fee, the common carrier (such as the telephone company) will guarantee that certain of these transmission problems will stay within a specified range considered acceptable for most modem designs. Thus you can do something about some of these transmission problems.

Possible transmission problems include:

- Attenuation distortion—frequency response
- Envelope delay distortion
- Nonlinear distortion
- Intermodulation distortion
- Frequency shift distortion
- Compandor distortion
- Signal-to-noise ratio
- Impulse noise
- Phase jitter
- Phase hits
- Gain hits
- Dropouts
- Echo

In the following subsections, we discuss seven of these transmission impairments divided into three groups. The first group is composed of impairments that are caused by transient electrical phenomena. Because of their transient and random nature, they are not really controllable with today's technology.

The second group includes two types of errors that the common carrier will limit to predefined ranges if you order what is called "C-conditioning" of your dedicated telephone line.

The third group consists of two other types of transmission errors that will be kept within predefined limits if you order "D-conditioning."

The rest of the problems listed are addressed by the common carrier on the basic dedicated line as long as the carrier knows the line is being used for data transmission. The common carrier gives no guarantees on the level of these other problems but historically has attempted to keep them at acceptable limits.

Transient Impairments

Three common transient impairments for which no practical permanent correction currently can be implemented by the telephone company are *phase hits, gain hits,* and *dropouts.*

Phase Hits These are sudden electrical forces that cause the phase of the signal to "jump" so that there is a difference in phase between the transmitted and the received signal.

Gain Hits A gain hit is a sudden increase or decrease in the level of the received signal that is less than 12 dB. Gain hits are especially bad for amplitude-modulated signals.

Dropouts A dropout is a decrease in the level of the received signal that is greater than twelve decibels and longer than four milliseconds. This phenomenon is especially hard on synchronous modems. It causes the loss of synchronization and forces the modems to go through a resync or a retrain cycle, thus reducing throughput.

These three types of transient errors are often called collectively *line hits* and are probably the most common source of errors in low-speed transmission. They are best compensated for by a good error-detection/correction scheme such as cyclic redundancy check (CRC) with go-back-N ARQ.

C-Conditioning

C-conditioning refers to the steps the telephone company takes to control attenuation distortion and envelope delay distortion. These steps normally include adding inductive coils called "loading coils" and various filtering equipment to the line.

C-conditioning is offered in five classes, C-1, C-2, C-3, C-4, and C-5. C-4 is the highest level of C-conditioning that can be obtained for ordinary private lines between customer sites and is the most expensive of these options. The specifications for your modems will indicate what level of conditioning, if any, is desired.

Some modems are designed to compensate for the types of distortion that C-conditioning corrects. An overcorrection (and, thus, distortion) can occur if C-conditioning is used with such modems. Be sure to follow the recommendations of your modem manufacturer.

Attenuation Distortion—Frequency Response As the signal generated by the modem travels down the telephone line, the attenuation of the signal is not uniform across all frequencies. In the voice telephone network, the attenuation is normally greater at higher frequencies than at lower frequencies. C-conditioning improves the high-frequency response and thus minimizes this distortion.

Envelope Delay Distortion As signals travel down a telephone line, they encounter resistance or delay. The amount of delay, like the amount of distortion, is frequency dependent; that is, the high-frequency components of a signal are delayed by a different amount than the low-frequency components. This difference in the amount of delay distorts the shape of the pulse.

A second problem caused by envelope delay at higher data speeds is to cause successively transmitted characters to overlap, causing what is called "intersymbol interference."

Since the human ear cannot detect these effects on speech, the voice network usually is not designed to correct these problems. C-conditioning provides the additional equipment necessary to keep envelope delay distortion within specified limits for data transmission.

D-Conditioning

D-conditioning addresses two problems: signal-to-noise ratio and harmonic or nonlinear distortion (also known as clipping). D-conditioning is primarily aimed at making a standard voice-grade private line suitable for use with 9600 bps modems.

Signal-to-Noise Ratio As its name implies, this is a measurement of the amount of noise on the line relative to the signal power. If the level of noise is too high, it can mask or distort elements of the signal and thus cause errors. This is not normally a problem with slower-speed modems, but with many 9600 bps modems the encoding techniques are so complex that very high signal-to-noise ratios are required.

Nonlinear Distortion *Clipping* occurs because the attenuation of the amplitude of a signal varies with its height. In other words, a high-voltage signal is attenuated more than a low-voltage signal, and the result is to flatten or "clip" the waveform. This is also called third harmonic distortion.

Because of the multiple levels used in coding 9600 bps signals, this nonlinear distortion makes it very difficult to distinguish between a different level (that is, a different bit combination) and a distorted signal. For this reason, D-conditioning is particularly important to anyone who plans to use 9600 bps modems.

Equalization

We discussed earlier the problems of envelope delay distortion and how it affects the shape of the received pulse and causes intersignal interference. The common carriers use devices called *equalizers* to compensate for envelope delay distortion. Modem manufacturers also include equalizer circuits in most conventional modems.

Equalizer circuits in modems help compensate for both envelope delay distortion and amplitude- and frequency-response distortion. Thus, with good modem equalization, C-conditioning may be unnecessary.

Equalizer circuits in modems generally fall into four classes: *manual, fixed, automatic,* and *adaptive.*

Manual Equalizers Manual equalizers can compensate for a range of distortion but must be adjusted by the user for the specific compensation desired for a particular line. They often have to be readjusted as conditions change; for example, a line can be rerouted, water can get into the line insulation or cable and change its characteristics, loading coils or other line treatments may be added or removed from the line—and all without the customer's knowledge.

Fixed Equalizers Fixed equalizers are permanently set at one compromise or average setting. They are also called *statistical equalizers* since the average setting is based on the statistical probability that it will be the best single setting for all lines.

Automatic Equalizers Automatic equalizers adjust themselves automatically to compensate for the specific distortion present on the line to which they are connected. Before the modem can transmit data, it must set the amount of equalization required. This takes place during the period between the modem's receiving the request-to-send (RTS) signal and sending back the clear-to-send (CTS) signal (known as the RTS/CTS delay). This period is called the "training period" or "training time."

Adaptive Equalizers Adaptive equalizers have the ability to sense changes to line conditions as they occur and thus constantly change their equalization parameters.

PROBLEMS WITH MODEMS ON DIAL-UP LINES

You will encounter slightly different types of problems when using modems on dial-up lines versus dedicated or private lines. In this section, we discuss some of the situations you will encounter with dial-up lines.

First, it is important to remember that, when using dial-up lines, you will probably be hooked up over a different physical circuit each time you dial up your connection. The routing of the call can also change frequently; for example, a call from San Francisco to Los Angeles might be routed via the coast route one time, via the valley the next time, and possibly via Chicago a third time. Your call might go on microwave one time and land lines another, or it may be all analog transmission once and the next time be converted to pulse coded modulation digital transmission. Thus the circuit conditions on a dial-up line can change considerably from one call to the next. Therefore, a good word of advice is, if the connection is poor, wait a few minutes and dial again.

Automatic Versus Fixed Equalization

Because of the changing nature of the dial-up connection, automatic equalization can be better than fixed or statistical equalization (although the latter is good for the average dial-up circuit).

Furthermore, when using the dial-up line in a half-duplex mode, you must remember that automatic equalization involves a training-time delay each time the line is turned around. In half duplex it is much better to use fixed equalization.

Echo Suppression

Because of the design of the voice telephone network, some of the transmitted signal energy is reflected back towards the transmitter from the receiving end. This causes echoes on the telephone line. To control echoes, devices known as *echo suppressors* are used.

When the dial-up network is used in a half-duplex mode, the echo suppressors must be removed from the line in the direction of the transmission. Unless a technique is used to keep the echo suppressors disabled at all times, they must be disabled each time the line is turned around. This adds around 400 milliseconds to the time required to turn around a half-duplex connection.

There are several approaches to overcoming this turnaround delay. One approach used in some modems is to send continuously signals that do not interfere with the data transmission but that keep the echo suppressor disabled. This is called a "fast-" or "quick-turnaround" modem.

Most of the requirements for quick turnaround usually come from the ARQ error-detection scheme because an acknowledgment or negative acknowledgment must be sent back to the transmitter after each message is received. Since most data are sent without errors, however, a reverse channel can be used for the ACK/NAK characters. This eliminates most of the need to turn the line around. Unfortunately, most computers do not have the special interface required to utilize a reverse channel.

Another approach is to use a go-back-N ARQ scheme. This reduces the number of direction reversals required and thus eliminates unnecessary delays for echo-suppressor turnaround and equalization.

The final approach is, of course, to utilize full-duplex modems on the dial-up line. Modems are available that can accomplish this at speeds of up to 4800 bps.

Automatic Dialing

Computer systems can automatically dial calls on the Direct Distance Dialing (DDD) or TWX networks and, when those calls are answered, transmit data. At the conclusion of the data transmission, the computer can terminate the call and go on to its next task.

To accomplish this automatic dialing, the computer typically utilizes a black box called an *automatic calling unit,* such as the Bell System's 801 or Racal-Vadic's VA811. The 801 auto dialer requires its own hardware interface, the RS-

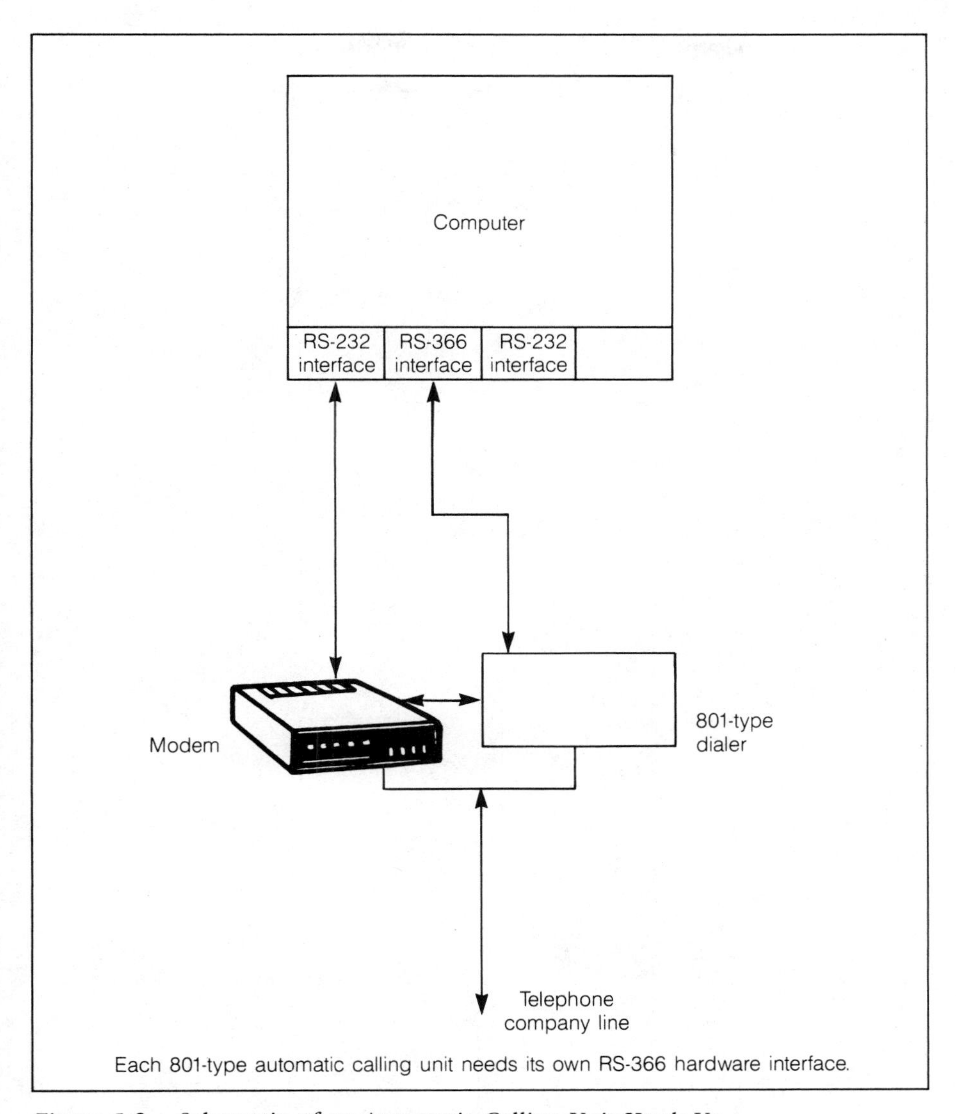

Figure 5-2. Schematic of an Automatic Calling Unit Hook Up

366 (see Figure 5-2). Adapters are available (for example, Racal-Vadic's VA831) that allow an 801 dialer to be connected to an RS-232 port (see Figure 5-3). As you can see from the figures, these automatic dialers hook up to the modem and occasionally may even be built into them.

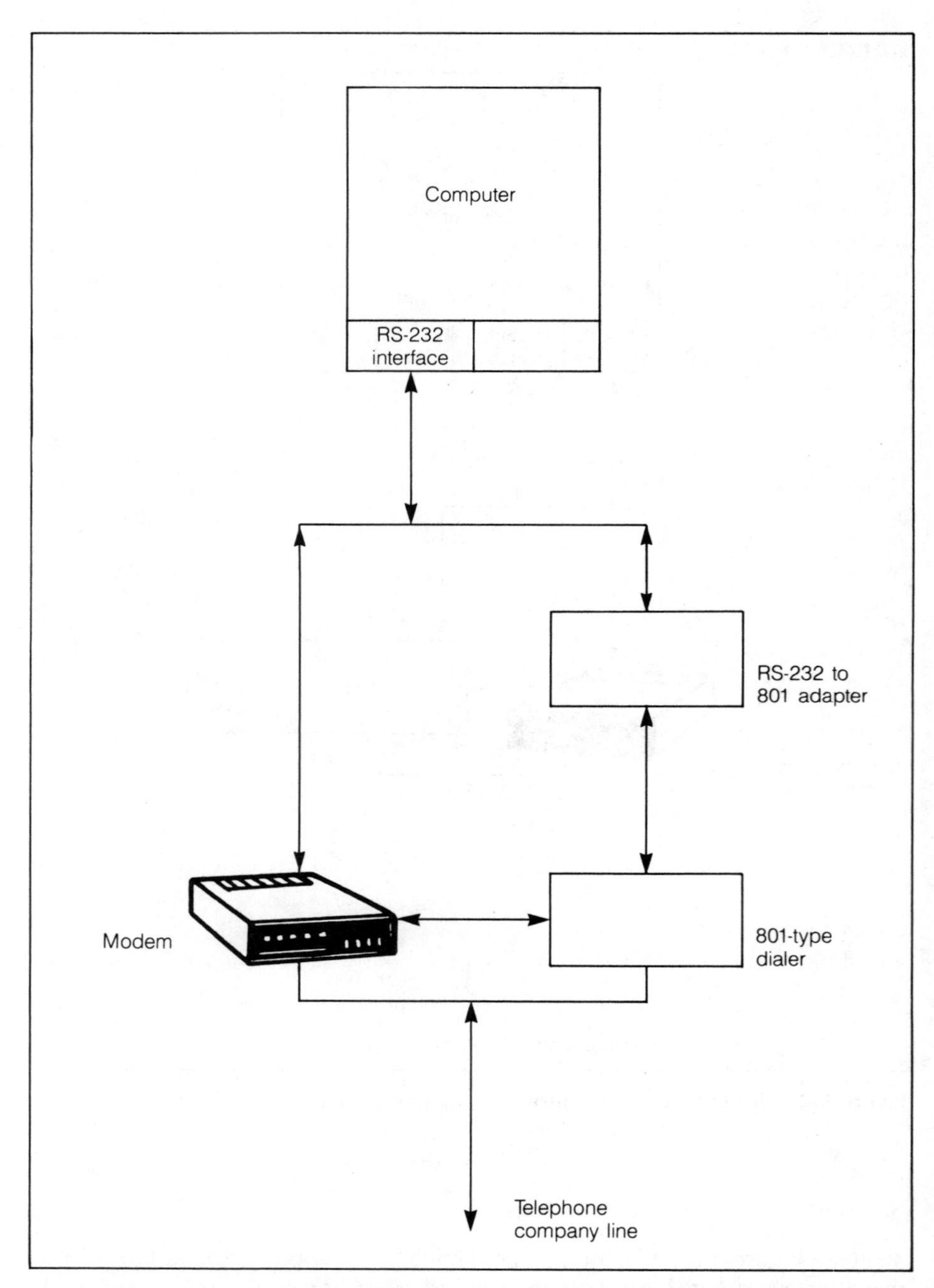

Figure 5-3. Automatic Calling Unit Using an RS-232 to RS-366 Adapter

MODEMS ON LEASED LINES

Modems that operate on leased lines differ somewhat from dial-up applications. First, leased lines can be ordered as four-wire lines that permit full-duplex operation at all speeds, so that line-turnaround time is eliminated. Second, conditioning can be ordered for leased lines (either C or D or both). Third, modems can operate at speeds of up to 9600 bps (or in some cases even higher) in a full-duplex mode. Fourth, leased lines tend not to be rerouted frequently; thus, line conditions that require transmission adjustments change much less frequently than on dial-up lines. Finally, leased lines are often multipoint lines.

Multipoint lines cause a special problem with equalization. If automatic equalization is required every time a different terminal is polled, then the net data throughput will be reduced considerably. One technique that is often used in this situation is called *forward equalization.* With forward equalization, each modem on the multipoint circuit is manually equalized for the conditions that prevail between it and the computer. Although this technique does require periodic readjustment, it eliminates the training time required by automatic equalization.

ASYNCHRONOUS-TO-SYNCHRONOUS CONVERSION

Let's say you wish to transmit data at 2400 bps or 4800 bps over dedicated circuits; however, the modems that transmit at this speed are synchronous, and your computer and terminals or other devices are all asynchronous. Luckily, there are black boxes that solve this problem.

One such black box is produced by ARK Electronic Products, Inc., and is called the elastic asynchronous-to-synchronous interface (EASI) or what I call an asynchronous-to-synchronous converter. Just plug one of these black boxes between your data terminal equipment and your modem (one is also needed at the other end of the circuit) and you have asynchronous DTE working with medium-speed synchronous modems (see Figure 5-4).

M/A COM and MICOM both offer a similar unit that includes the cyclic redundancy check for error control. An asynchronous-to-synchronous converter/error controller built into a synchronous modem is also available. You can also buy an asynchronous-to-synchronous converter/error controller that does full-duplex-to-half-duplex conversion for use with 2400 bps or 4800 bps half-duplex dial-up modems.

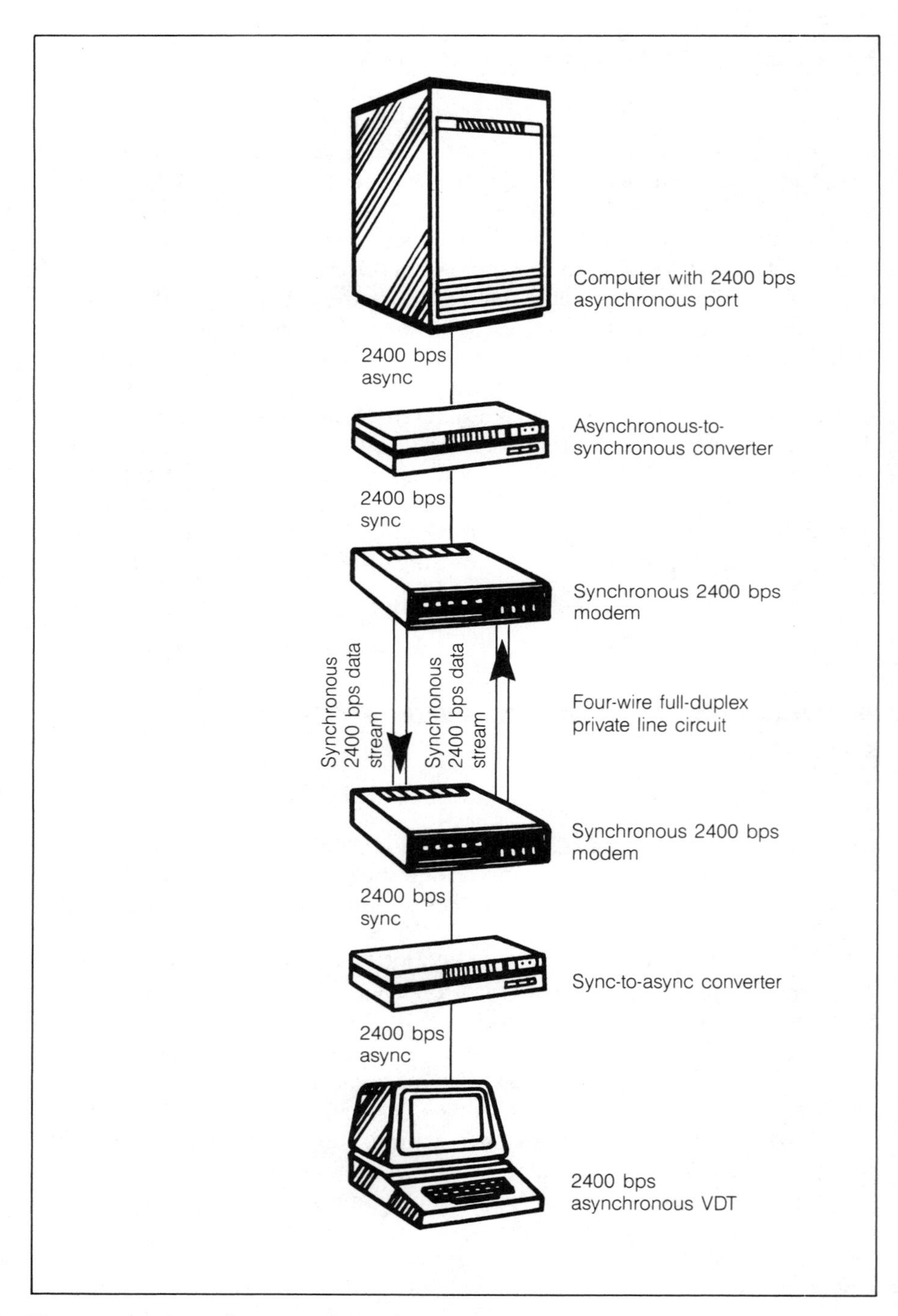

Figure 5-4. Asynchronous-to-Synchronous Conversion

MULTIPORT MODEMS

Multiport modems combine a multiplexor and a modem and thus enable the user to split a modem's data stream so that it can accommodate two or more slower DTE's. (These modems are also called split-stream modems.) This allows you to reduce the cost of data communications lines.

Figure 5-5 shows an example of four terminals each operating at 1200 bps that are connected to one multiport modem. The modem combines the data streams from the four terminals into one 4800 bps data stream and transmits it to the modem at the other end of the line, where the data stream is again broken into four separate data streams that are fed to their separate ports on the computer.

The multiport modem gives the user the benefits of a multiplexor (which will be discussed in Chapter 8) without having to buy a separate black box. Typically, the multiport modem costs more than a conventional modem but less than a modem plus a multiplexor.

TRIPLE AND QUADRUPLE MODEMS

Triple and *quadruple modems* pack the equivalent of three or four different modems in the same black box. They typically include enough intelligence to decide what speed and type of modem is at the other end and to adjust themselves automatically in order to transmit signals at the highest compatible speed. Usually this black box is actually only one modem that can run in any one of three or four modes or protocols but not in more than one at the same time.

In 1977 Racal-Vadic introduced the VA3467 triple modem that included a VA3400 (Vadic's 1200 bps full-duplex dial-up modem), compatible modem (the Bell Systems' 1200 bps full-duplex dial-up modem), and a Bell 103 compatible modem (the Bell Systems' 300 bps full-duplex modem). In 1982, Racal-Vadic introduced its VA4400 quadruple modem, which incorporates its 2400 bps full-duplex dial-up modem, its VA3400, a Bell compatible 212A modem, and a Bell compatible 103/113 modem.

Since the separate 2400 bps modem only costs about 10 percent less than the quad modem, the quad modem is a wise investment if a computer port is going to be accessed by different terminals each with different modems (see Figure 5-6). In Figure 5-6, any one of the four VDT's (A, B, C, or D) can dial the computer's auto-answer telephone line, and the quad modem will automatically go into the correct mode to establish communications with the modem on the calling VDT.

Another useful combination of modems is found in Datatronix's DATAport II. This is designed for Telex, TWX, and electronic-mail applications. The DATAport II comes with up to four types of ports: a Bell 101C compatible port for TWX; a Bell 103J compatible port for DDD 0 to 300 bps transmissions; a Western Union TIM-81-1(F1/F2) port compatible for Telex; and an RS-232C port to allow

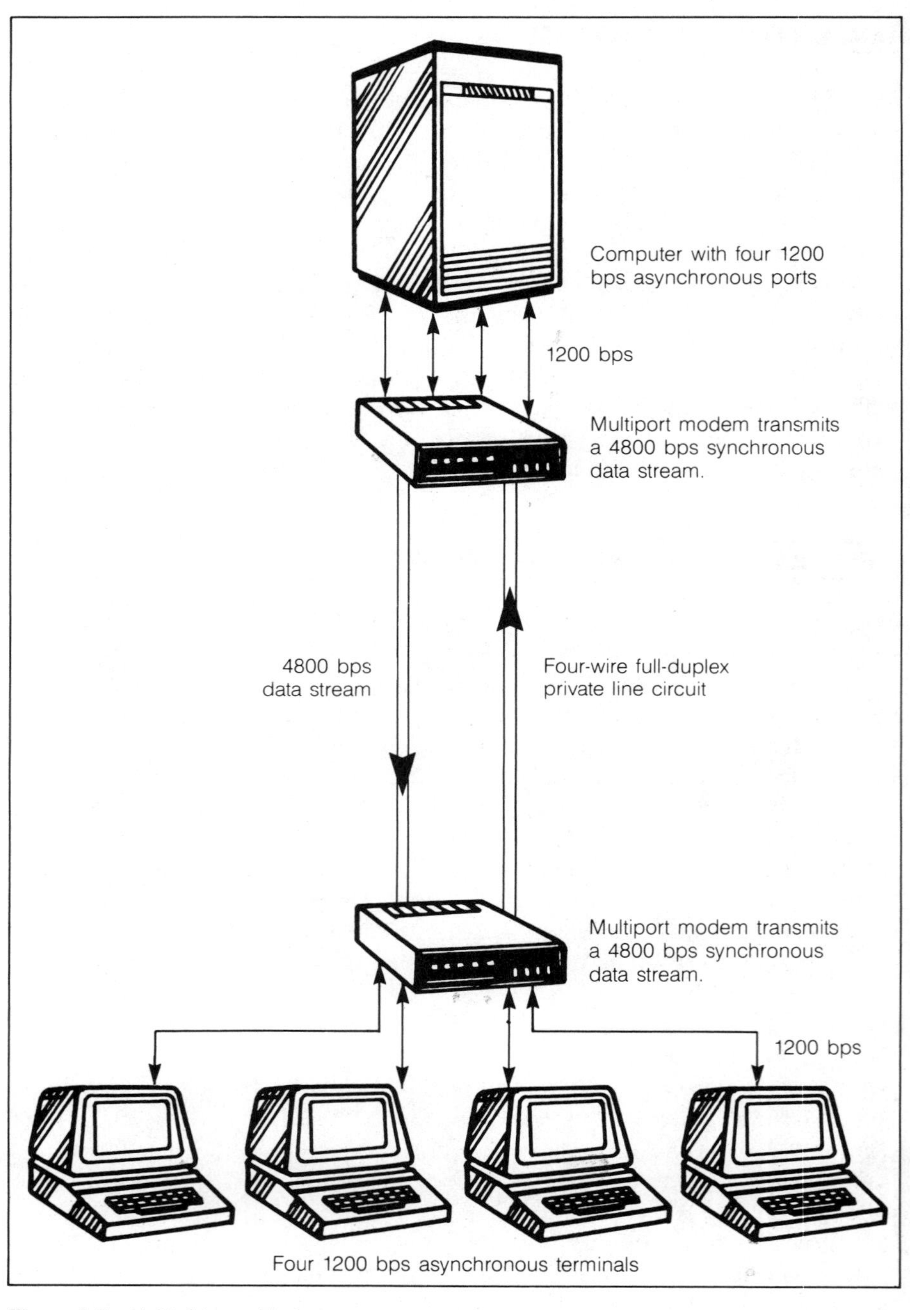

Figure 5-5. A Multiport Modem

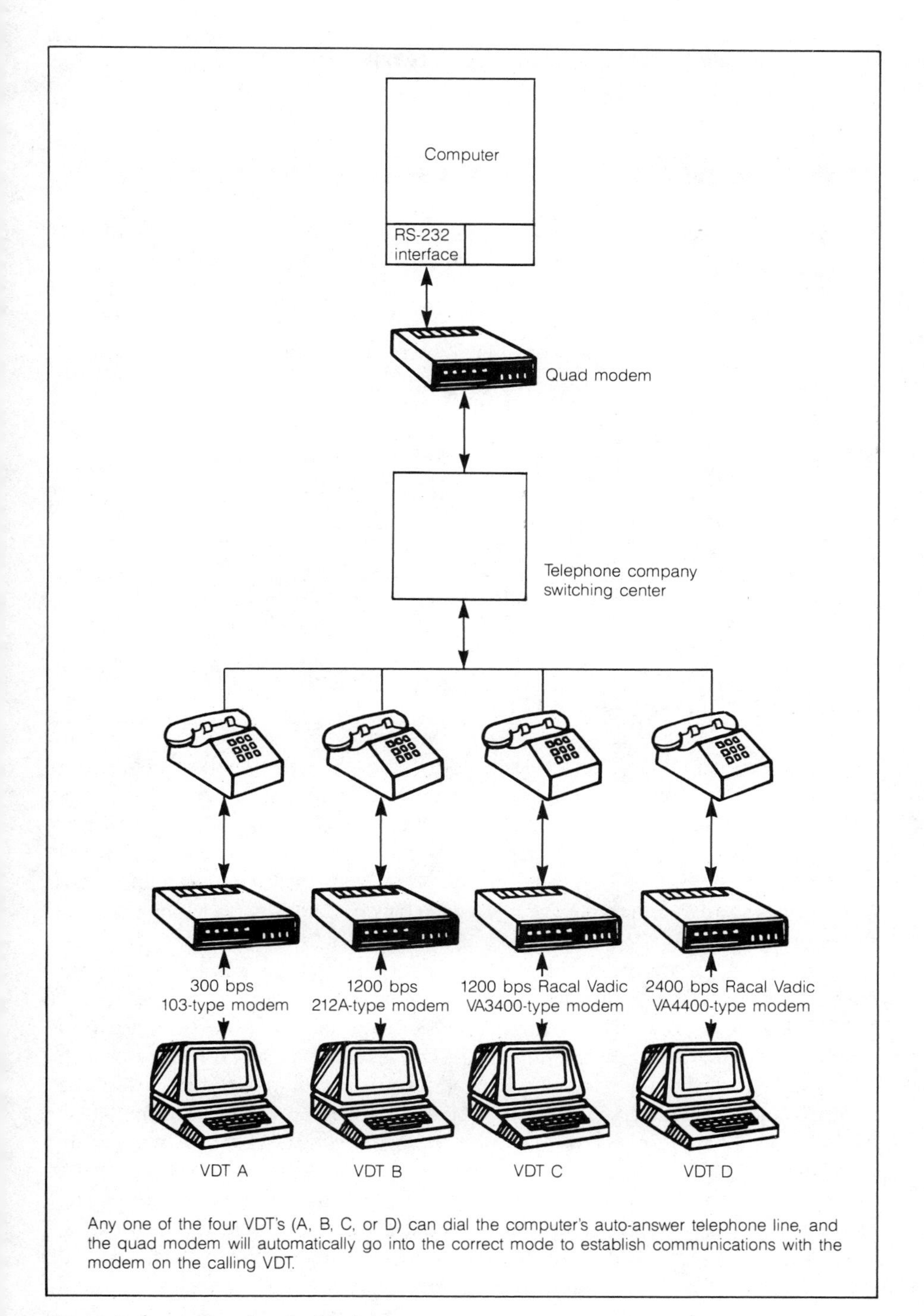

Any one of the four VDT's (A, B, C, or D) can dial the computer's auto-answer telephone line, and the quad modem will automatically go into the correct mode to establish communications with the modem on the calling VDT.

Figure 5-6. A Quadruple Modem

connection to higher-speed modems for electronic-mail applications. The DATA-port II also has auto-answer and auto-dial capabilities.

TYPICAL MODEMS AVAILABLE FROM AT&T

The following chart lists the most popular modems available from AT&T and their distinguishing characteristics.

NAME	SPEED (IN BPS)	MODE	TYPE OF LINE
103/113	300	Async FDX	Dial-up or leased line
202	1200	Async HDX	Dial-up
202	1800	Async HDX	Leased line with C-conditioning
201	2400	Sync HDX	Dial-up or 2-wire leased line
201	2400	Sync FDX	4-wire leased line
208	4800	Sync HDX	Dial-up or 2-wire leased line
208	4800	Sync FDX	4-wire leased line
209	9600	Sync FDX	4-wire leased line
212	1200	Sync FDX	Dial-up or 2-wire leased lines
212	1200	Async FDX	Dial-up or 2-wire leased lines

POINTS TO REMEMBER

Modems come in many different sizes and shapes. You've now had a chance to become familiar with the wide range of choices confronting you in selecting modems and related "black boxes." This chapter has given you enough information to ask your vendors many questions about their products. Find out what is currently available from the different vendors and decide which features are most appropriate to your specific application before deciding on which modems to get.

One excellent source for "black boxes" of all sorts that can help you solve many of your data communications interfacing and interconnection problems is the Black Box Corporation in Pittsburgh, Pennsylvania. This corporation pub-ishes a catalogue called the *Blackbox Catalogue* that is a very useful reference.

One last word of caution: when installing your modems be very careful that they are set up or "strapped" correctly. Most modems have many options. These options have to be selected or "strapped" before the modem can be used. Many users have installed modems and found they did not work, only to discover that the modems were strapped incompatibly or incorrectly. Read the modem installation manual very carefully and doublecheck all straps or switch settings. If, after you have done all that, the modems are still malfunctioning, check with your supplier's (or the modem manufacturer's) customer-service department. This department is usually more than willing to assist you.

6

NETWORKS—TOPOLOGIES AND ACCESS METHODS

If you want to connect terminals or other peripherals to your computer, you need to consider what kind of network to use. A *network* consists of the transmission system (such as wires, cables, satellites, and so on) and the associated control software and hardware (I call this software and hardware combination the *network control equipment*) that are used to connect computers and peripherals or other computers to each other. Although networks usually connect devices that are in different locations, the wiring and control equipment used to connect peripherals to a computer in the same room can also be called a network.

Most of the examples in this chapter relate to networks using telephone company voice lines or circuits to connect equipment in different cities. But the basic concepts presented apply to other types of networks as well.

Different manufacturers have their own communication schemes, which are often not compatible with those of other manufacturers. These individual schemes typically specify the topology (or physical structure) of the network, the network access technique and the protocol that will be used. This chapter explains network topology and access methods (Chapter 7 will tell you more about protocols) so that you can determine which communication scheme best suits your application *before* you lock yourself into one manufacturer's networking requirements.

THE FIVE NETWORK TOPOLOGIES

One method of segregating networks is based on their topology (*topology* here means the physical configuration of the computer and its peripherals in the network). This section looks at the five major topologies for communications networks: the star, the mesh, the bus, the tree, and the ring.

The Star or Point-to-Point Network

In the *star network* (often referred to as a *point-to-point network*), the computer is at the center of the network and each peripheral has its own communications line that connects it to this central computer. In some star networks, a PBX replaces the computer at the center of the network. (A PBX, or private branch exchange, remember, is an automated telephone system used by many businesses that is characterized by the use of a central operator and by the user dialing 9 to get an outside line.)

Looking at Figure 6-1, you can see the star shape of this topology. The telephone network, for example, is based on the star topology, with each home or business having a telephone line (the local loop) that connects it with the central office (see Figure 6-2).

The Cluster Controller and the Star Network In a widely used approach to connect terminals and printers to a central mainframe computer, a network is configured with "stations," each one of which involves several devices. The

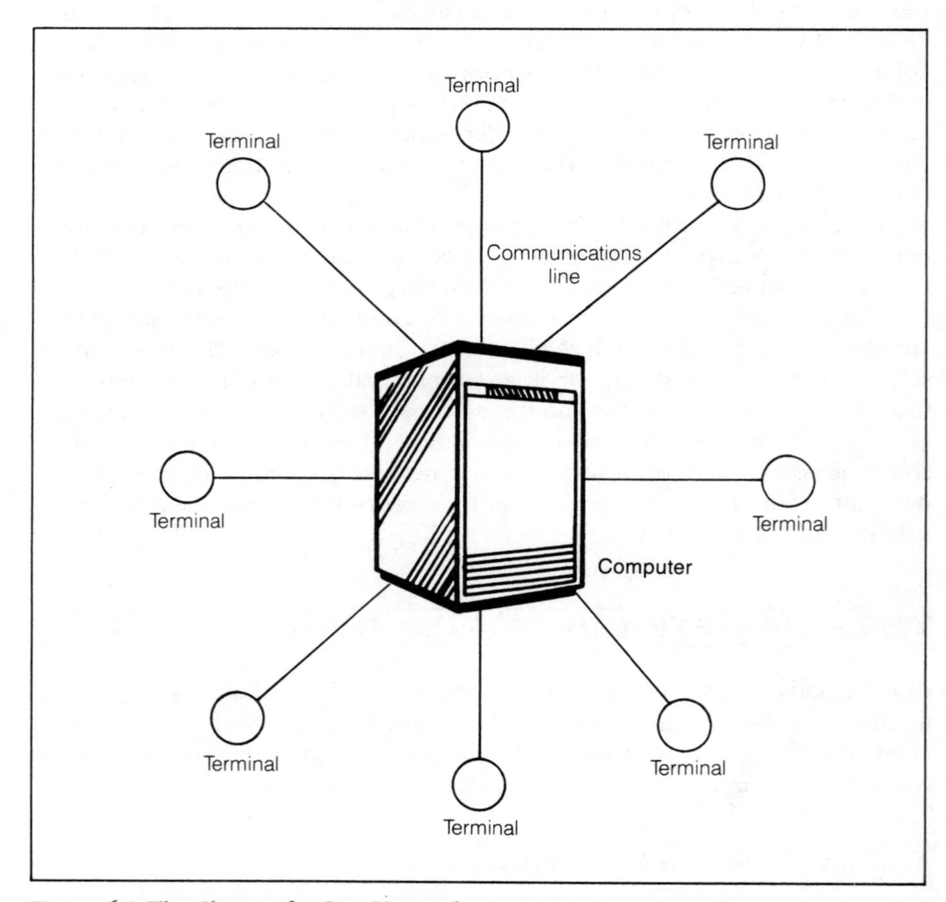

Figure 6.1 The Shape of a Star Network

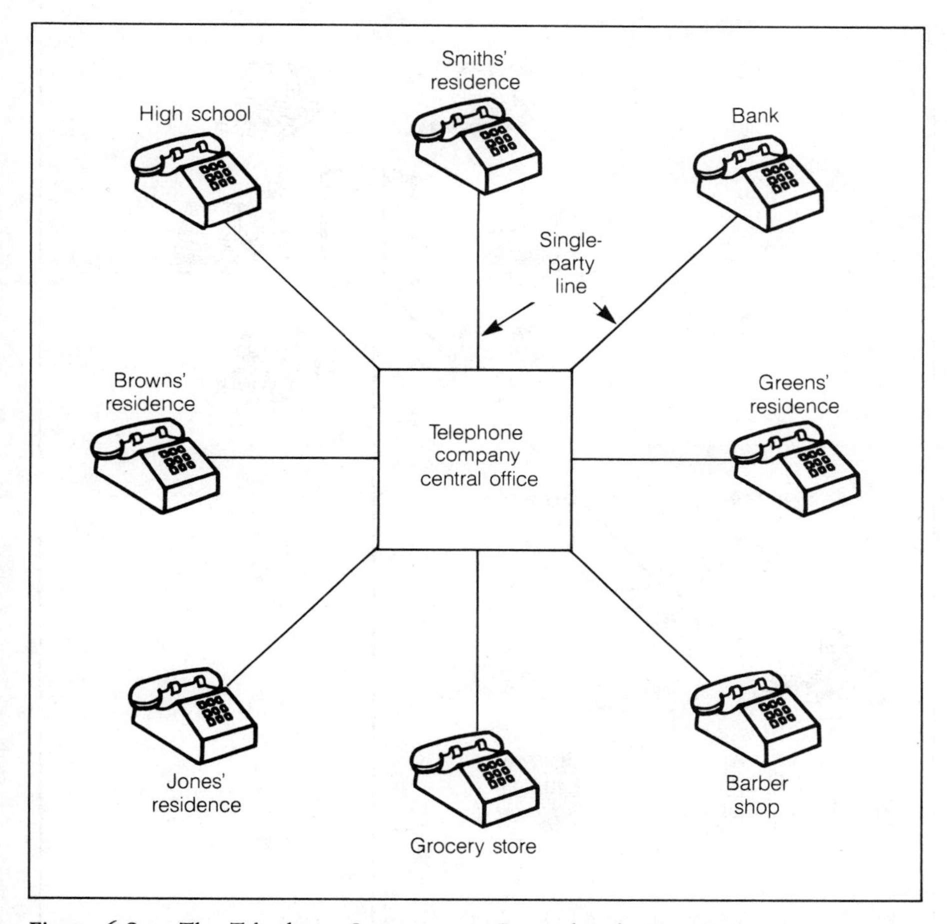

Figure 6-2. The Telephone System as an Example of a Star Network

station, then, can consist of a "terminal cluster controller" with terminals or printers attached to it. A *cluster controller* is the hardware and associated software that consolidates some of the data communications functions that would otherwise have to be duplicated in each one of the devices connected to the network.

The traffic from a single cluster controller and the many devices attached to it can be of such a great volume that one communications line can handle only one cluster controller. In this case, it is economical to use the star topology to connect each station to the mainframe computer.

Multiple Star Networks It is possible to connect two or more star networks together as shown in Figure 6-3 to form a larger, more complex network referred to as a *multiple star network.* Message-switching systems are often set up as

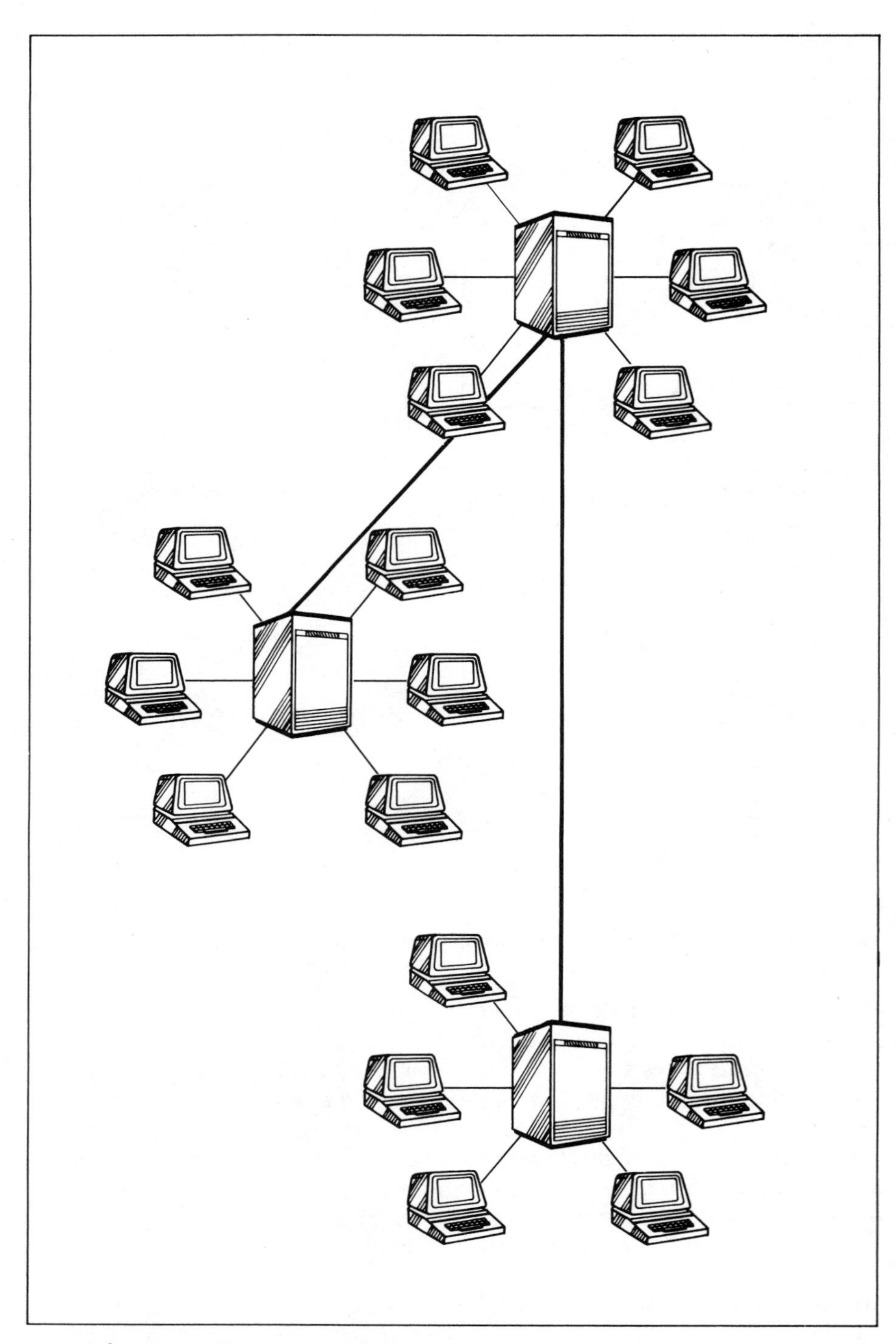

Figure 6-3. A Multiple Star Network

multiple star networks, and the multiple-city telephone networks used by many large companies as well as the nationwide Direct Distance Dialing (DDD) voice telephone network also use the multiple star network topology. In fact, star networks that use cluster controllers are actually examples of multiple star networks, since the peripherals form a star around each cluster controller, and the cluster controllers form a star around the host computer.

The Simplest but the Most Expensive Historically, the point-to-point or star network has been the most common network used with mini- and microcomputer systems because it requires the simplest and least expensive of the network control equipment used by the four topologies. However, when lightly to moderately used peripherals are connected one per line, the star topology is usually the most expensive in terms of communications line costs, especially if long-distance communication is involved.

The reason that the star topology is so expensive when used to connect individual terminals directly to a central computer can be seen by looking at Figures 6-4 and 6-5. Figure 6-4 shows three remote terminals in each of two cities connected to a computer over 3,000 miles away and four local (that is, in the same location as the computer) terminals connected to the same computer.

Notice that each terminal has its own line to the computer in a point-to-point network (remember, there are only two points of connection, one at each end, for every line). The cost of a separate line for each terminal is not usually expensive for the local terminals, but to connect the six terminals in California to the computer in New York requires six transcontinental circuits. If dedicated, full-time circuits are used, this can be very expensive.

A *dedicated* circuit is a circuit for the exclusive use of a particular customer; it is often referred to as a "leased" line or circuit since dedicated lines are usually "leased" from a common carrier such as AT&T. Because these dedicated lines can be very expensive and because there is often no need to use the lines many hours every day, standard dial-up voice telephone lines are frequently used in point-to-point data networks.

Another reason that the star network can be expensive is that each peripheral requires a separate port on the computer, since each line must terminate in its own computer port (see Figure 6-5). An additional problem is that some computers may be limited as to the number of ports available.

Because of these drawbacks to the star network and because the cost of network control equipment continues to decrease dramatically, other topologies are gaining popularity with mini- and microcomputer systems.

The Mesh Network

In the star network, a remote device cannot send a message directly to another remote device; rather, it must be sent to the central computer where the message can then be relayed to the destination device. In the *mesh network,* on the other

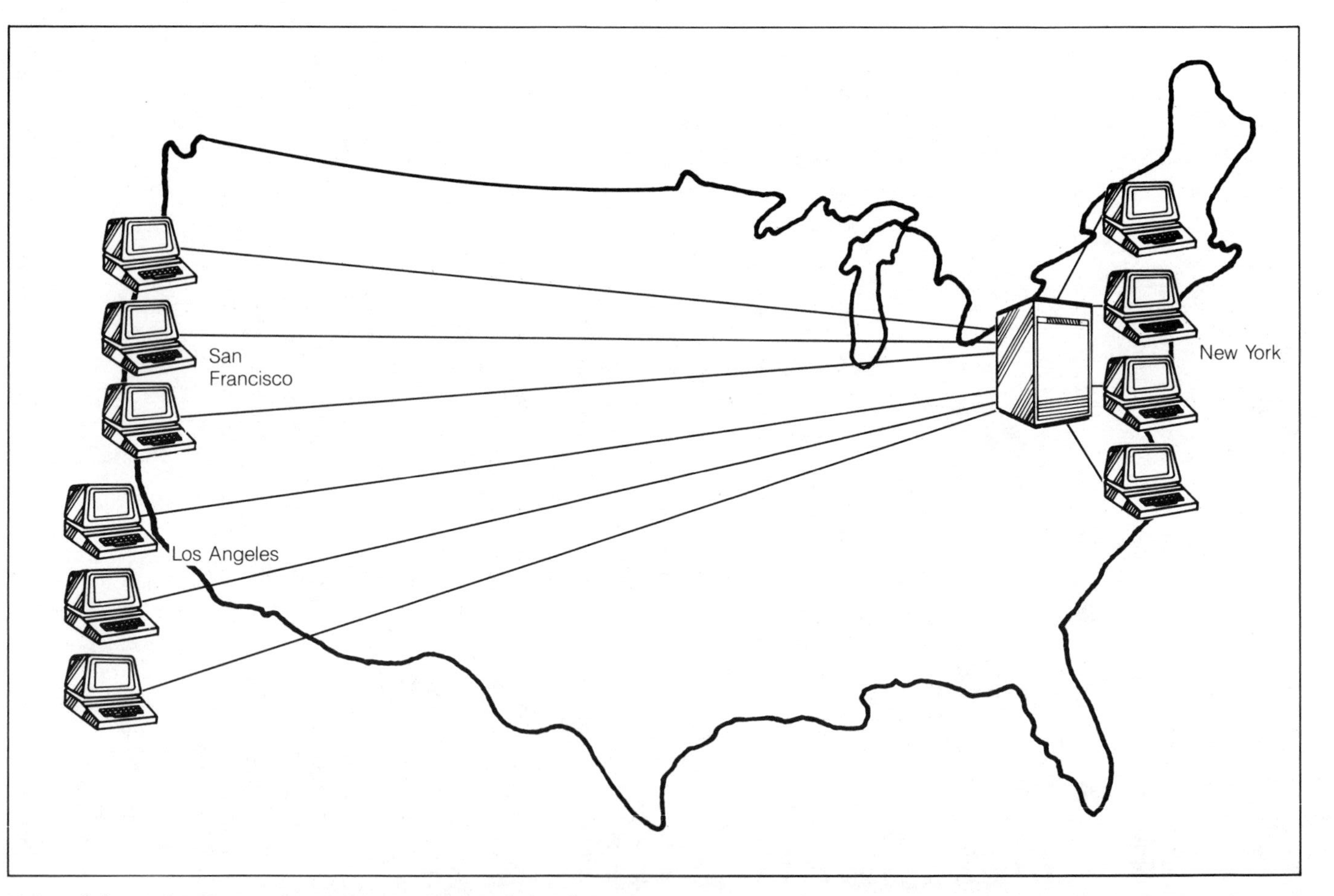

Figure 6-4. A Star Network Connecting Terminals in Three Cities to a Central Computer

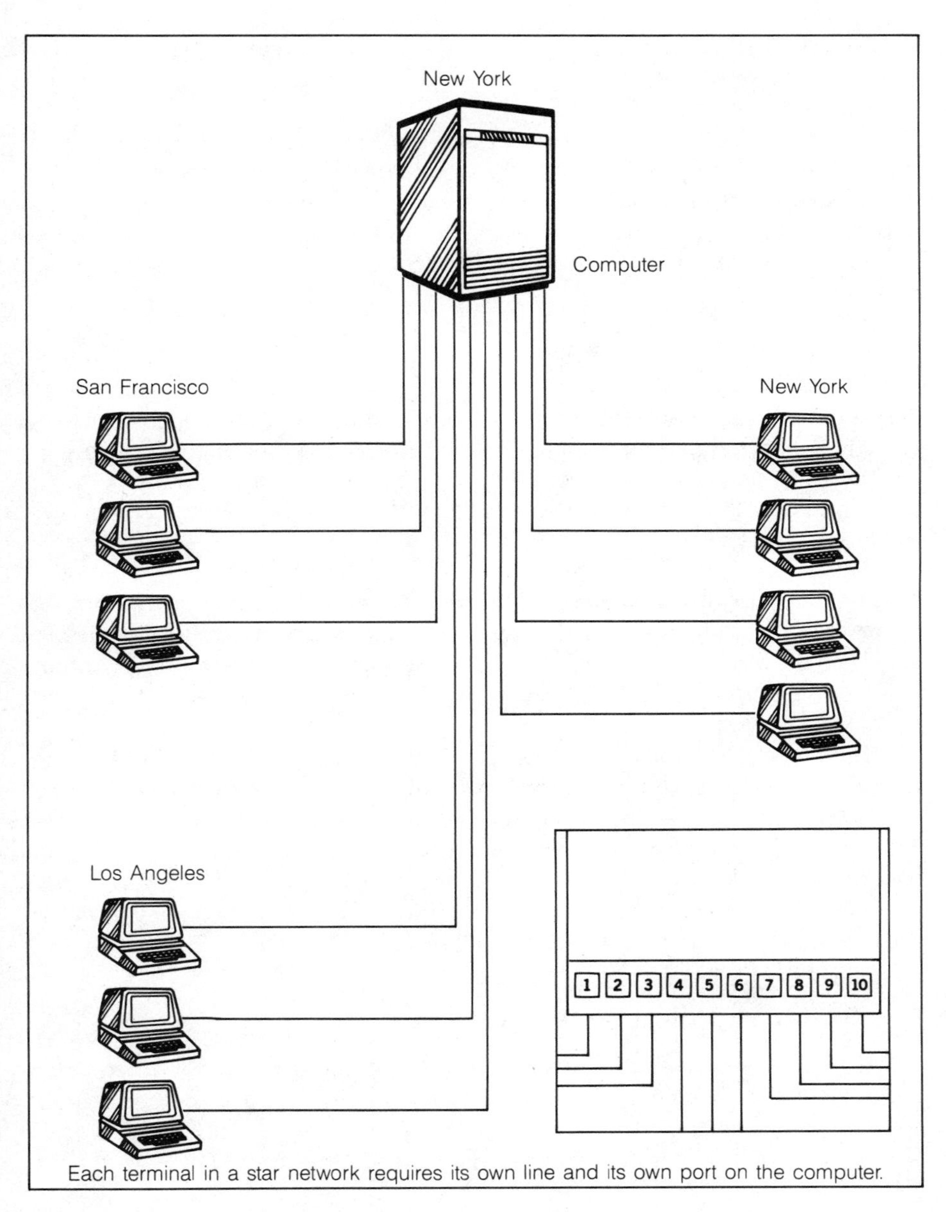

Each terminal in a star network requires its own line and its own port on the computer.

Figure 6-5. Why a Star Network Can Be Expensive

hand, all devices can send messages to other devices. The messages do not have to be relayed by a central control unit but may be relayed by an intermediate node on a multilink path.

The points in the mesh where two lines are connected together are referred to as *nodes*. The nodes consist of the network control equipment to which the user's device is connected. Mesh networks are characterized by multiple paths between any two users on the network. A network that connects several computers together is a good example of a system that might make use of a mesh network.

There are basically two types of mesh networks—the fully connected mesh and the partially connected mesh.

The Fully Connected Mesh In the fully connected mesh, each device is directly connected to every other device by a point-to-point line (see Figure 6-6). The fully connected mesh network can obviously reach a point where the number of devices that have to be connected with every other device just becomes too large to be practical.

The Partially Connected Mesh The partially connected mesh provides direct routes between only some of the devices. If a message is sent to a device to which no direct route is available, then the message must be relayed by another node (see Figure 6-7). Usually at least two alternate routes are provided to and from every device.

The routing control and the storage capability required at each node make the nodes very expensive, and, thus, partially connected mesh networks have not been employed frequently. However, the partially connected mesh now forms the basis for a complex type of network called the "intelligent communications network." This network is exemplified by IBM's Systems Network Architecture (SNA) and DEC's DECNET and by packet-switching networks such as the GTE Telenet and ARPANET, which will be discussed in Chapter 7.

The Bus or Multidrop Network

In the bus network (often referred to as a multidrop or multipoint network), the computer is typically at one end of a long line or bus, and the peripherals are attached or "dropped" at various points along the line. The computer can also be at a point on the bus rather than at an end (see Figure 6-8).

Advantages of a Bus Network The bus or multidrop network saves both communications lines and computer ports. These advantages can be seen in Figures 6-9 and 6-10, which show the terminals from Figure 6-4 connected in a bus network.

Figure 6-9 shows the computer connected to the end of two different bus networks. In this configuration, two computer ports are used, one for the Cali-

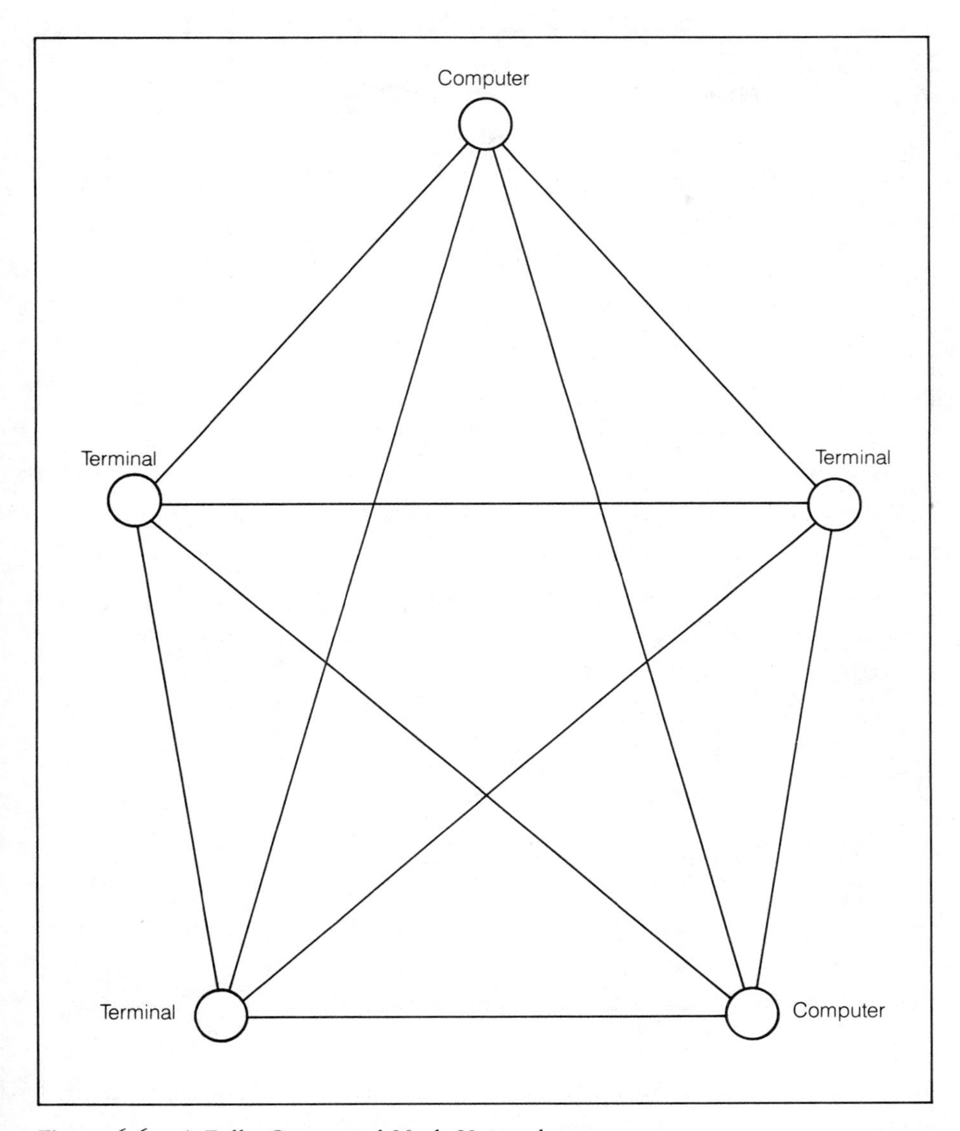

Figure 6-6. A Fully Connected Mesh Network

fornia line and one for the local New York line. See that the six lines to California needed in the star network are replaced in the bus network with only one line that runs to Los Angeles and then up to San Francisco.

Figure 6-10 shows the computer connected to the center of a single-bus network. In this configuration, only one computer port is used. Although this single-bus network saves a computer port, it puts more terminals on the same

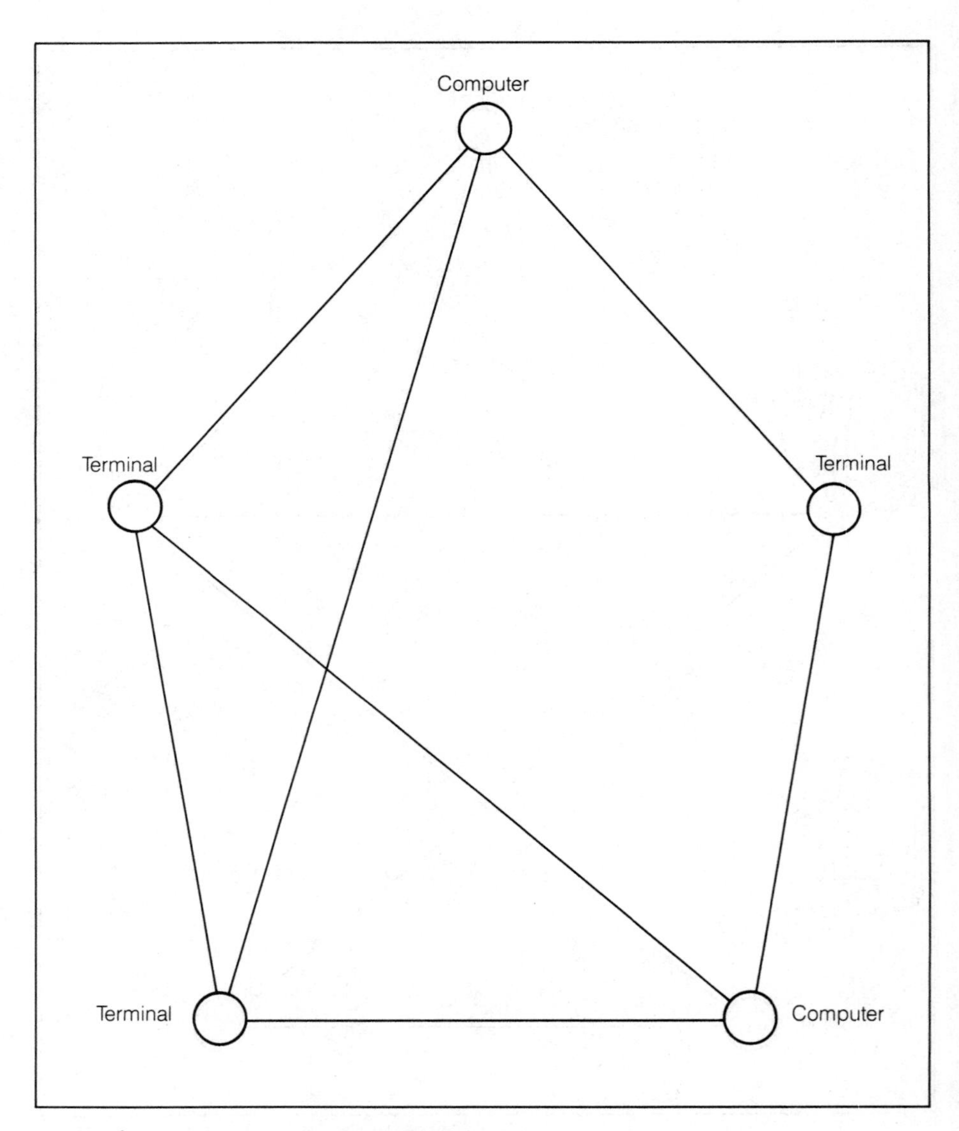

Figure 6-7. A Partially Connected Mesh Network

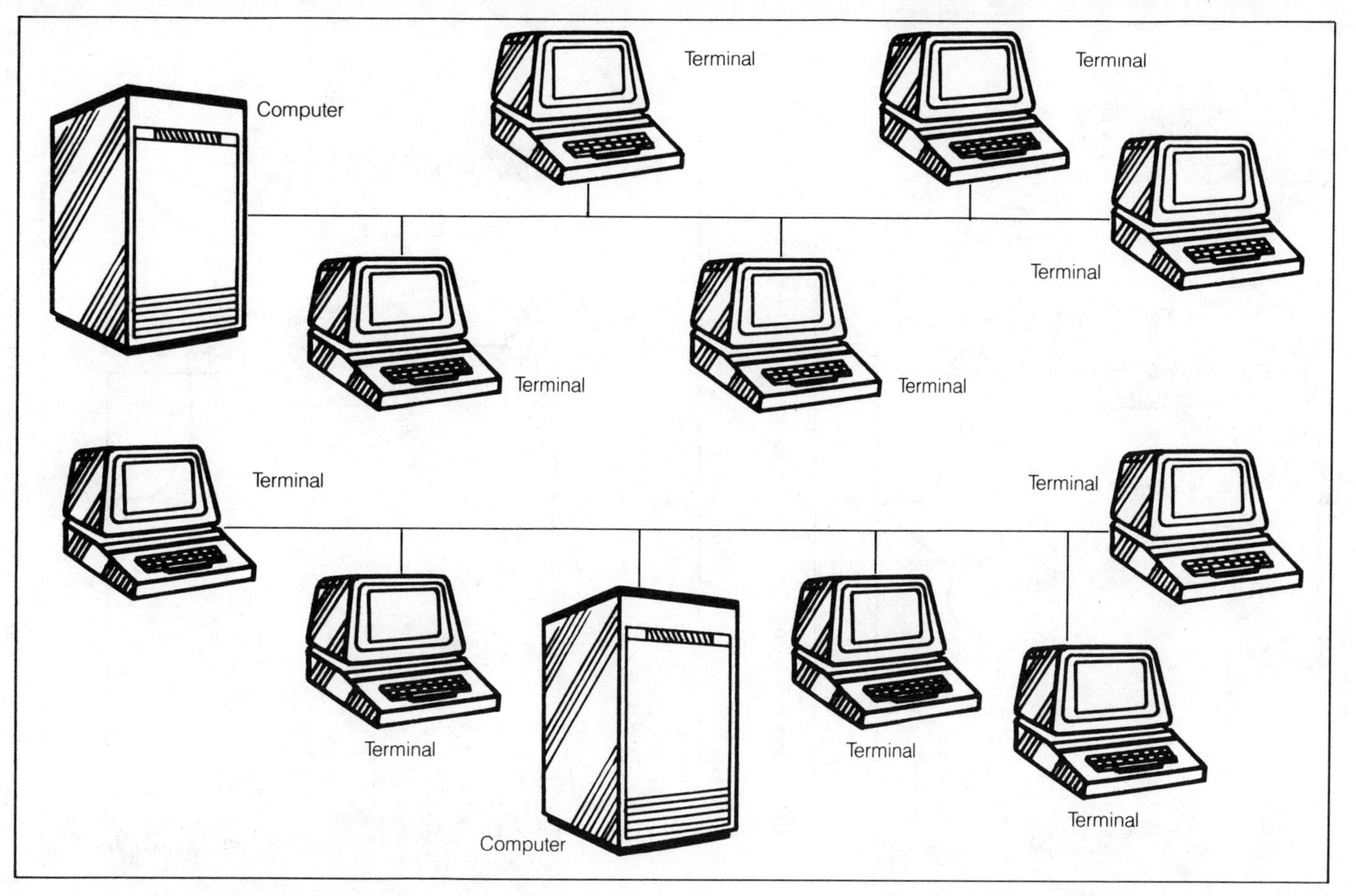

Figure 6-8. Two Examples of the Bus Topology

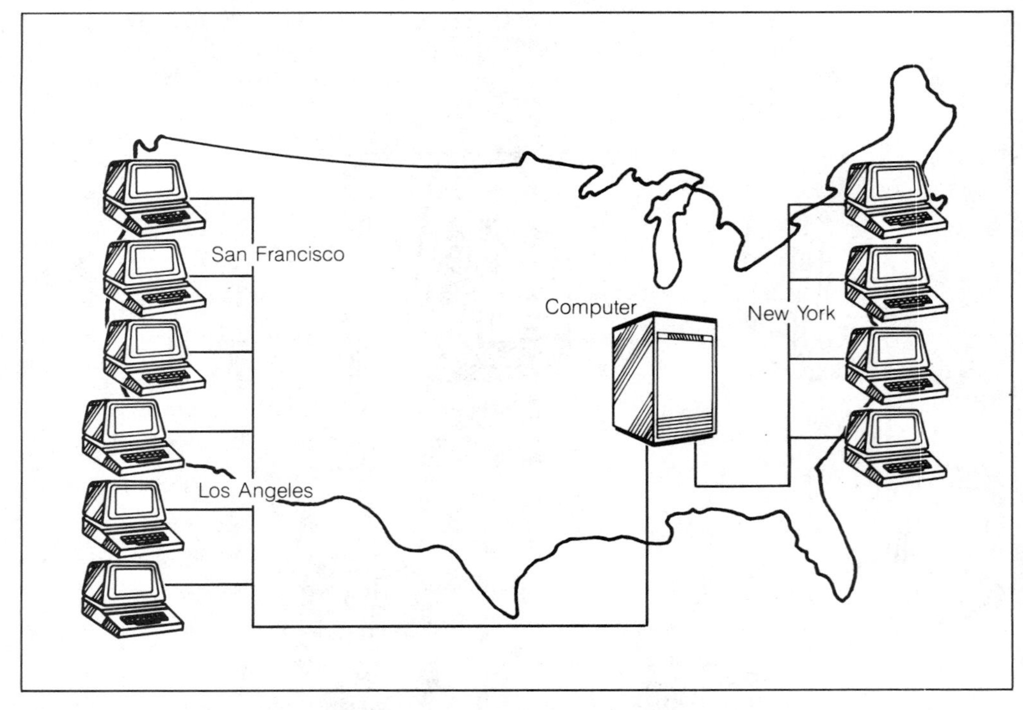

Figure 6-9. A Bus Network Using Only Two Lines and Two Ports

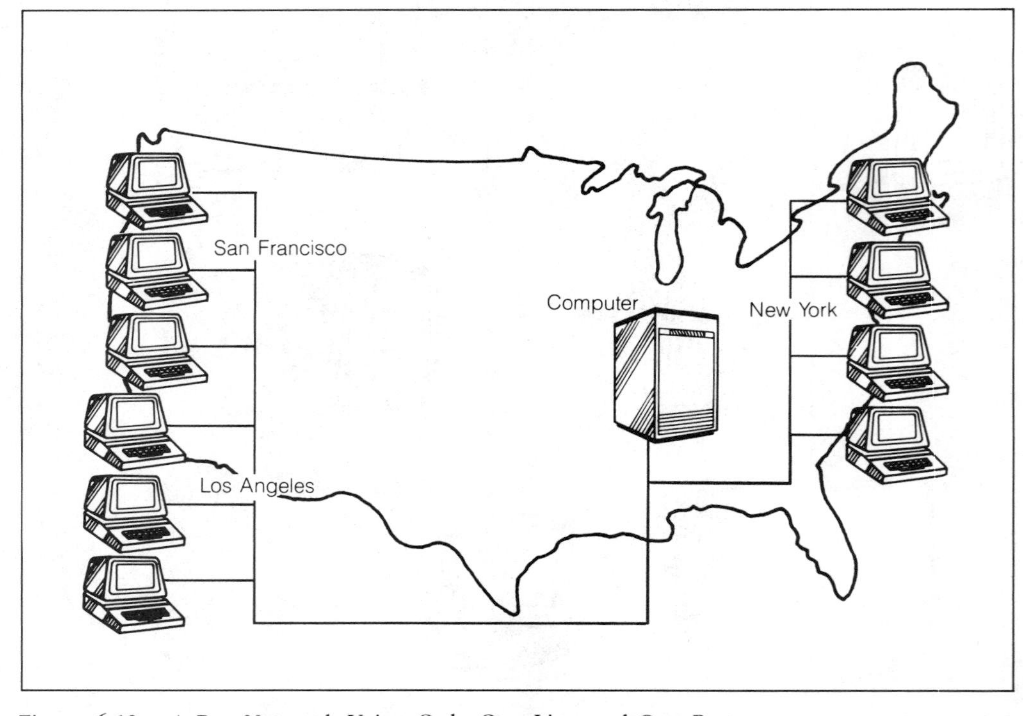

Figure 6-10. A Bus Network Using Only One Line and One Port

line and same port. Therefore the traffic on that line and port will be heavier, and you will need to analyze whether the response time will be satisfactory.

Also note that there are ten terminals on the one line. Some computer systems will not be able to address this many terminals on one line, in which case you would have to split the network into two lines as shown in Figure 6-9.

Splitting the one circuit into two not only solves the addressing problem but also provides other benefits. If the local terminals are split off onto a separate circuit, this local circuit will be inexpensive. Overall response time on both circuits is improved since there is less traffic on each circuit. Overall "up" time is also improved; in a one-bus network, if the one line or the one computer port fails, all terminals on that line will be down, but in a two-bus network, the entire network does not fail if one line goes down.

Drawbacks of a Bus Network In a basic star network, only one device talks over a line to the computer, and each computer port has only one peripheral connected to it. However, in the bus or multidrop network, the main line and its one computer port must handle all of the communications to and from the computer for all of the peripherals on that bus. Thus, the network control equipment must be much more complicated in order to keep all of the communications from getting scrambled or from interfering with each other. In addition, the data-carrying capacity of the bus must be large enough to handle the communications load of all devices combined rather than just the load of one device.

Thus, with the bus network, computer port costs and line costs are significantly lower than with a star network in most configurations, but the network control equipment costs are higher. And the response time of the devices on the bus may be slower, particularly if it is limited by the data-carrying capacity of the bus.

The Tree Network

When two or more star or bus networks are connected together at one of their nodes, they produce a topology called a *tree.* Note that if the star networks are joined by connecting two central points, we call the resulting network a multiple star.

If the buses of two bus networks are joined, the result is one longer bus. If three or more bus networks are connected together at the same point (that point has the topological appearance of a node but does not have to have the intelligence to connect devices to it), the result is still a tree.

The tree topology, like the bus topology, is a multidrop or multipoint network.

If a tree structure has a clearly defined base node as shown in Figure 6-11 (node X), then the tree is called a "rooted tree." If there is no clearly defined base node (as in Figure 6-12), then it is called an "unrooted tree."

The bus, star, and tree networks could all be combined on one large multidrop system resulting in a *mixed topology* (see Figure 6-13). Other topologies can also be mixed to form a network such as a mesh of trees (see Figure 6-14).

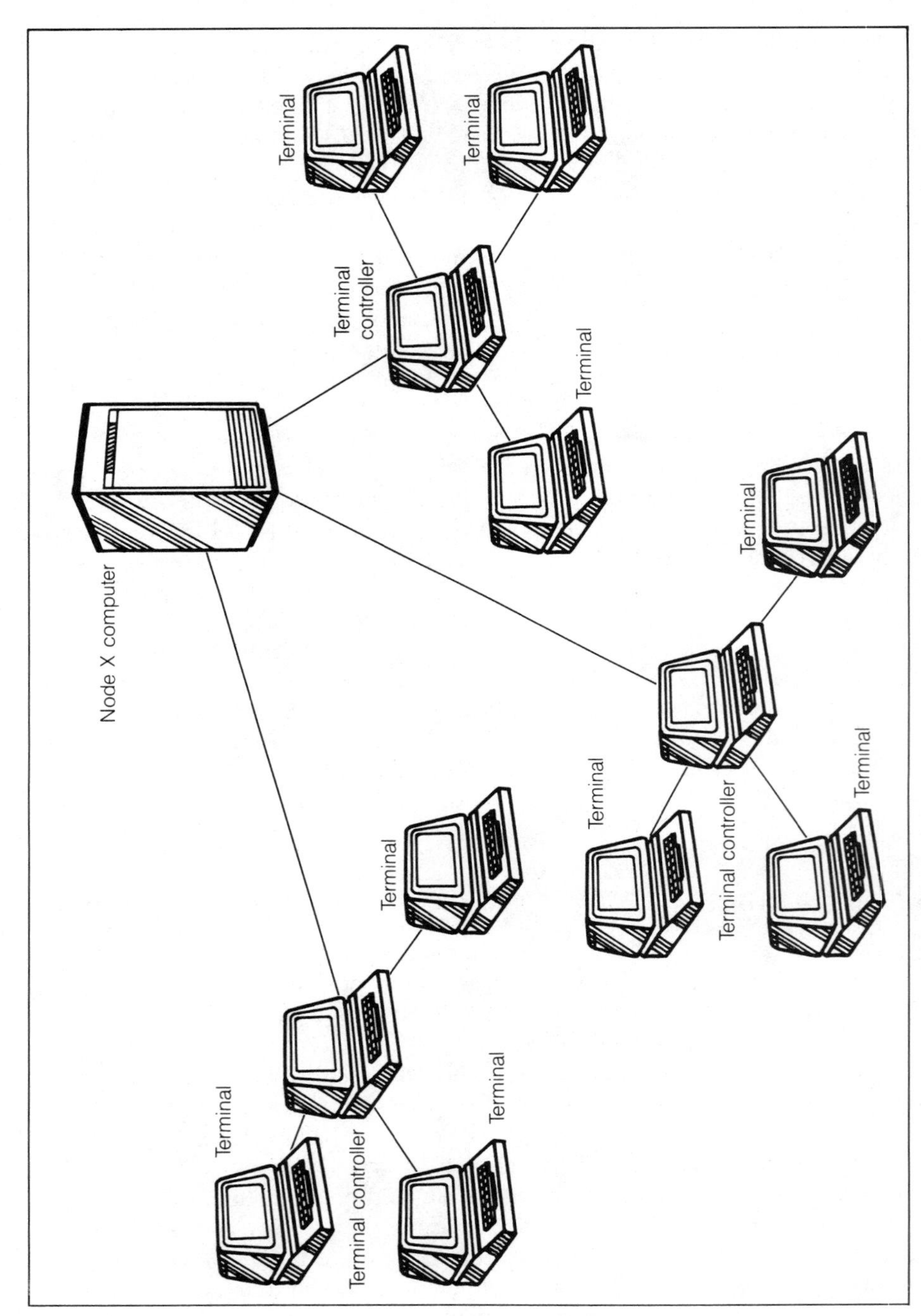

Figure 6-11. A Rooted Tree Network Formed from Three Star Networks

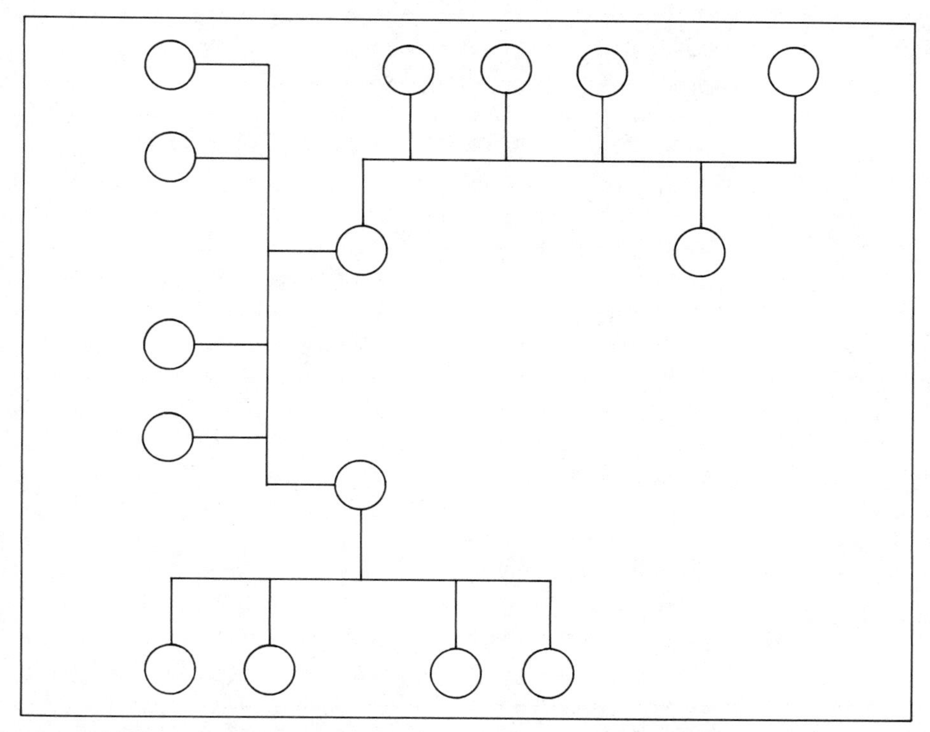

Figure 6-12. An Unrooted Tree Network Formed from Three Bus Networks

The Ring or Loop Network

The fifth type of network topology is the *ring* or *loop*. The ring or loop network consists of a series of nodes joined together to form a circle or a ring. However, the ring or loop network is *not* simply a bus network with a communications line that runs continuously through each node. In the ring or loop network, the communications line terminates at each node and starts again at the other side. In order for a message to go to a node and then continue on to the next node, the message must be repeated or retransmitted by the first node (see Figure 6-15).

The ring or loop network, then, is actually composed of a series of point-to-point circuit segments connecting adjacent nodes on the ring or loop; yet it is usually thought of as a multipoint network, since a number of nodes all receive and transmit their messages over the same series of point-to-point circuit segments.

Rings Versus Loops or Buses If one of the nodes in the ring is a controlling station (such as the central computer), then the ring is usually referred to as a *loop*.

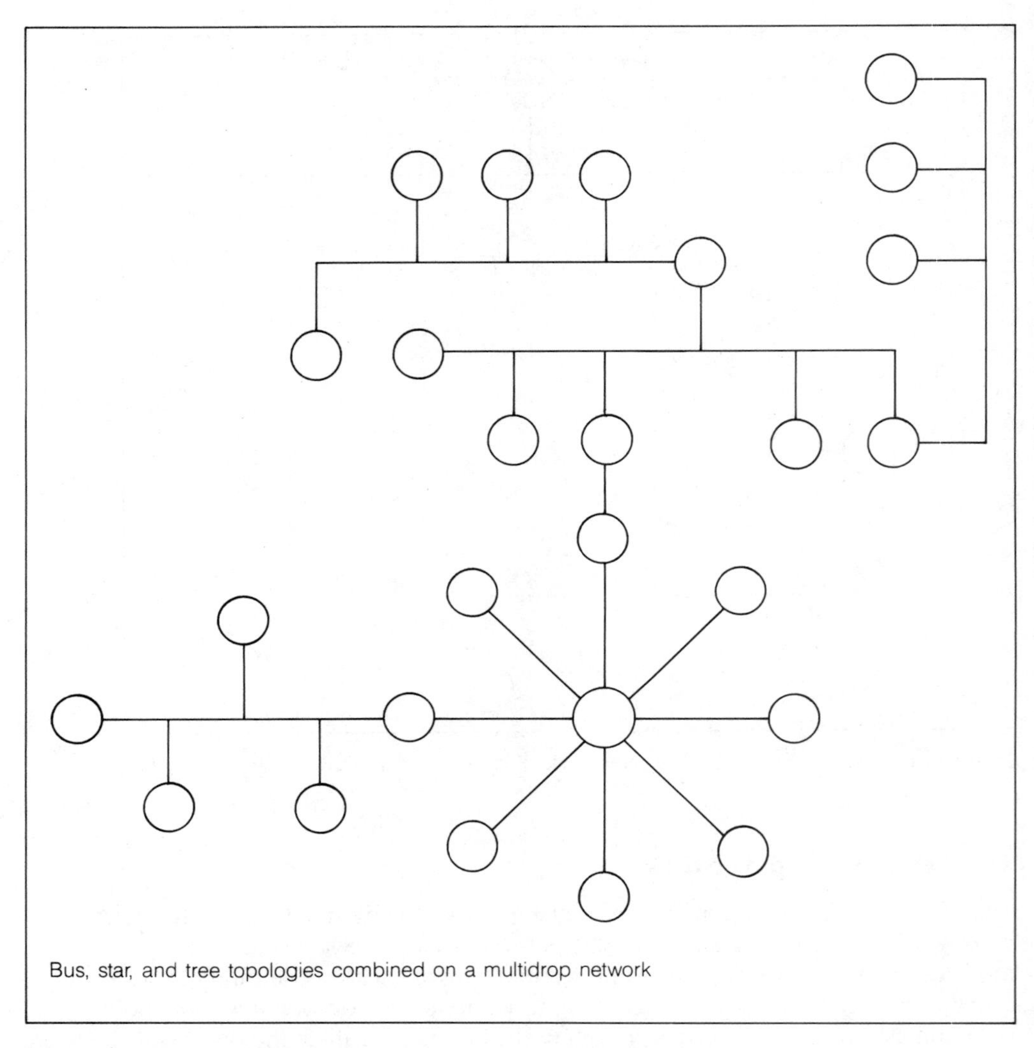
Bus, star, and tree topologies combined on a multidrop network

Figure 6-13. Mixed Topology

Sometimes the term ring will be applied to a bus network that has both ends of the bus joined together. The difference is that, on a ring or loop, the message flow is interrupted at each node and retransmitted by that node to the next node in the direction of flow. In the normal bus topology, the message flows along the bus and can be heard by all nodes in the direction of the flow. Note that before the message is retransmitted from one node to the next node along the ring or loop, the message may be altered by the node. Thus, on a ring or loop network, all nodes do not necessarily hear the same message.

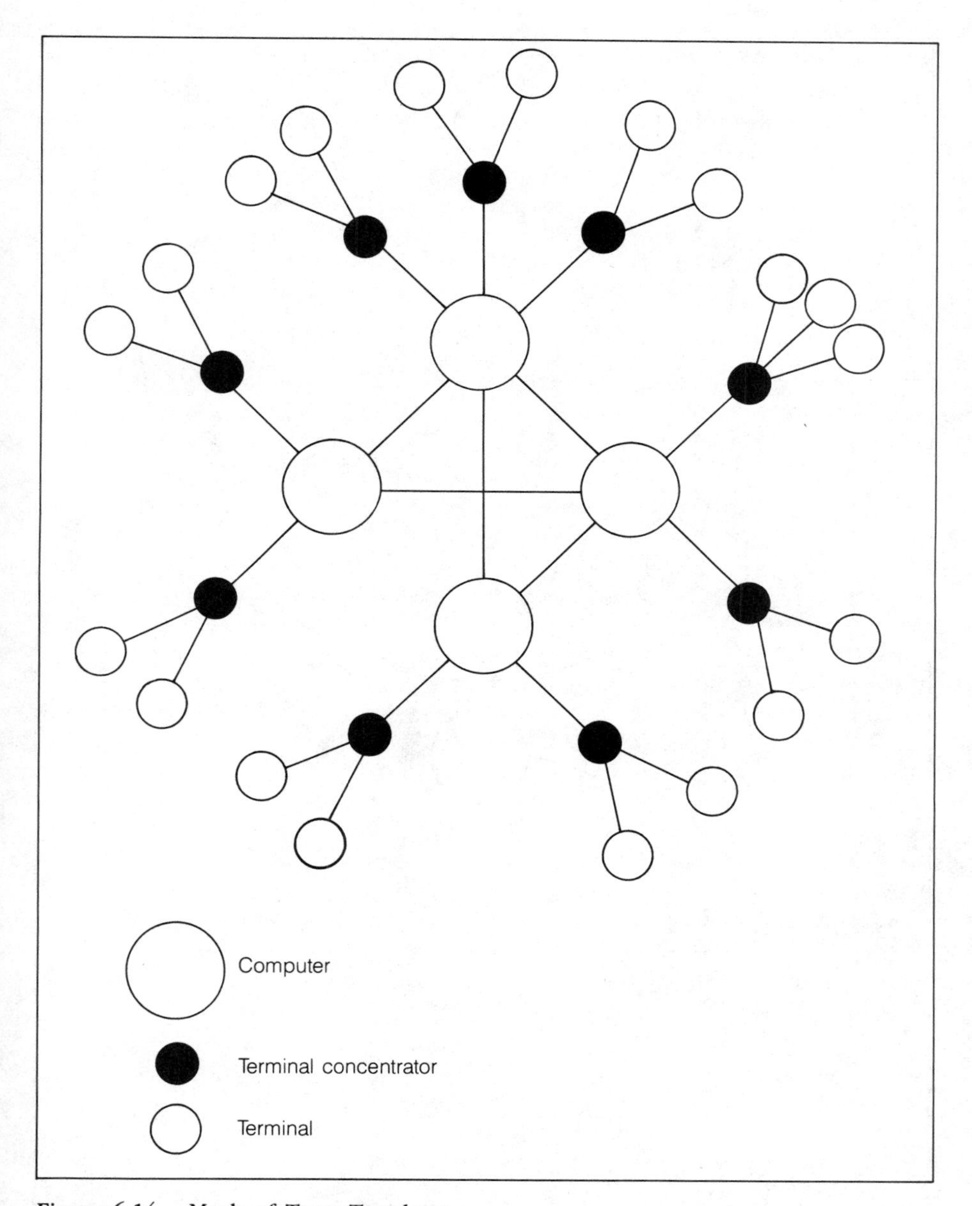

Figure 6-14. Mesh of Trees Topology

Direction of Flow One major difference you should keep in mind between the ring or loop network and the star, mesh, bus, and tree networks concerns direction of flow. In the star, mesh, bus, and tree networks, communications must flow in both directions—that is, from the computer to the terminal and

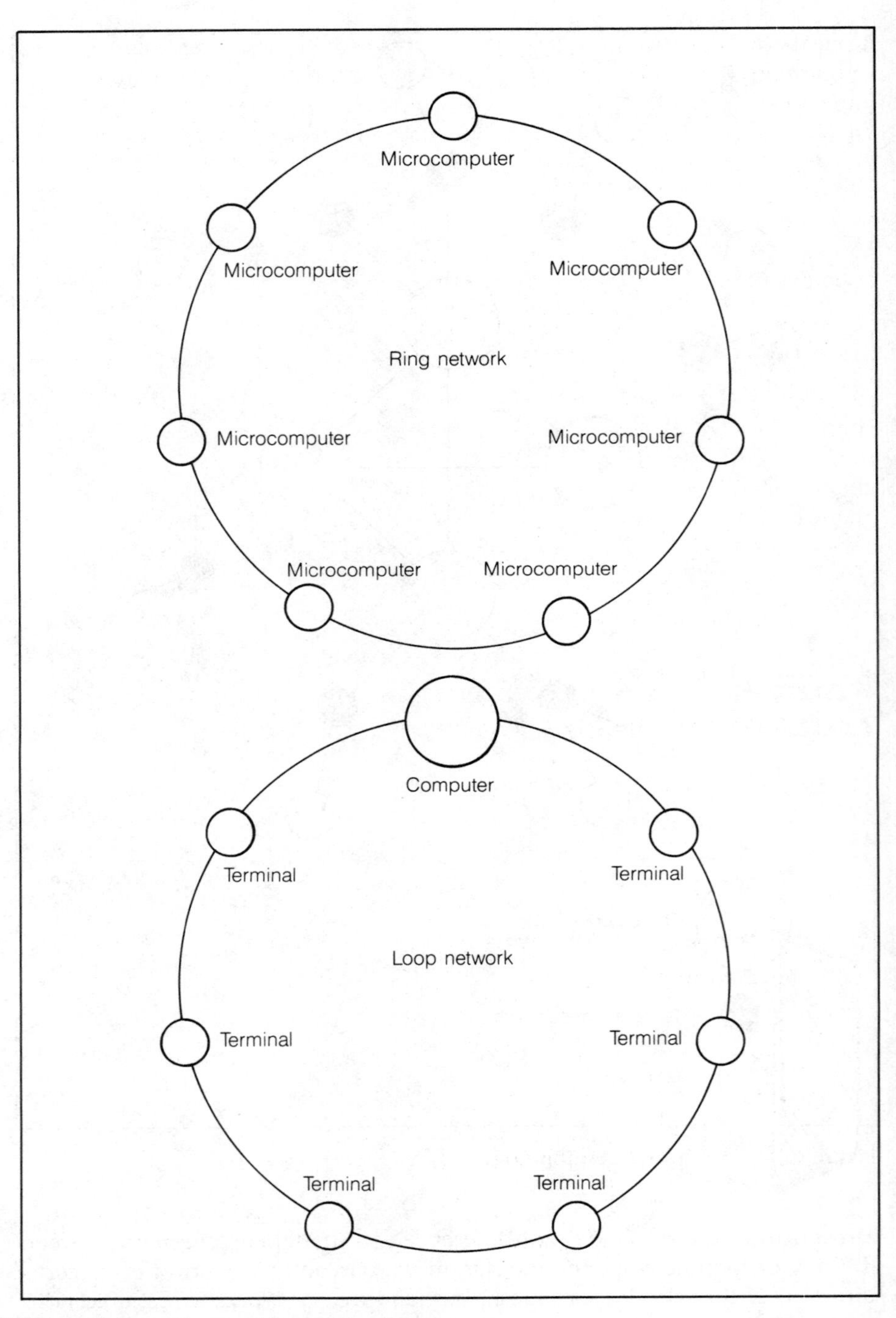

Figure 6-15. Ring and Loop Topologies

from the terminal to the computer (see Figures 6-16 and 6-17). In the usual application of the ring or loop network, communications can flow in only one direction (see Figure 6-18).

In some implementations of the ring or loop topology, however, full-duplex circuits are used, and communications does take place in both directions around the ring or loop.

Figure 6-19 shows what a loop topology would look like in our three-city example. With the loop topology, two lines to California are needed, since the loop must be complete, whereas one line suffices in a bus network.

Gateways

You need to know one other term here that is related to network topologies: the connecting link between two separate networks (along with its associated communications control software and hardware) is called a *gateway* between the two networks. The gateway may be as simple as a telephone circuit linking the two networks, or it may include complex hardware and software to allow messages to move between two otherwise incompatible networks. If the networks are incompatible, the gateway may supply code, protocol, and speed conversion among other functions.

FOUR ACCESS METHODS FOR MULTIDROP NETWORKS

One of the negative aspects of multidrop networks is that the messages to all of the devices travel over the same multidrop line, and thus complicated network

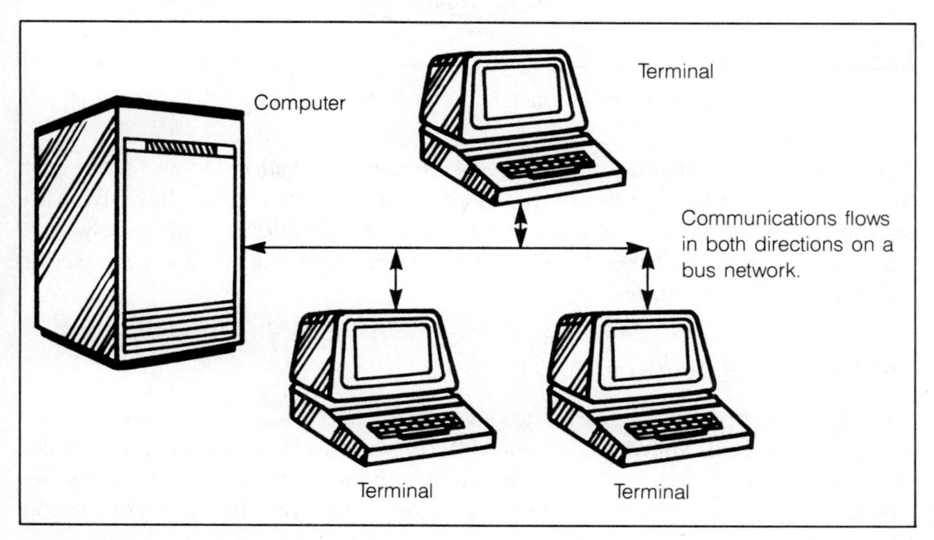

Figure 6-16. Direction of Flow on a Bus Network.

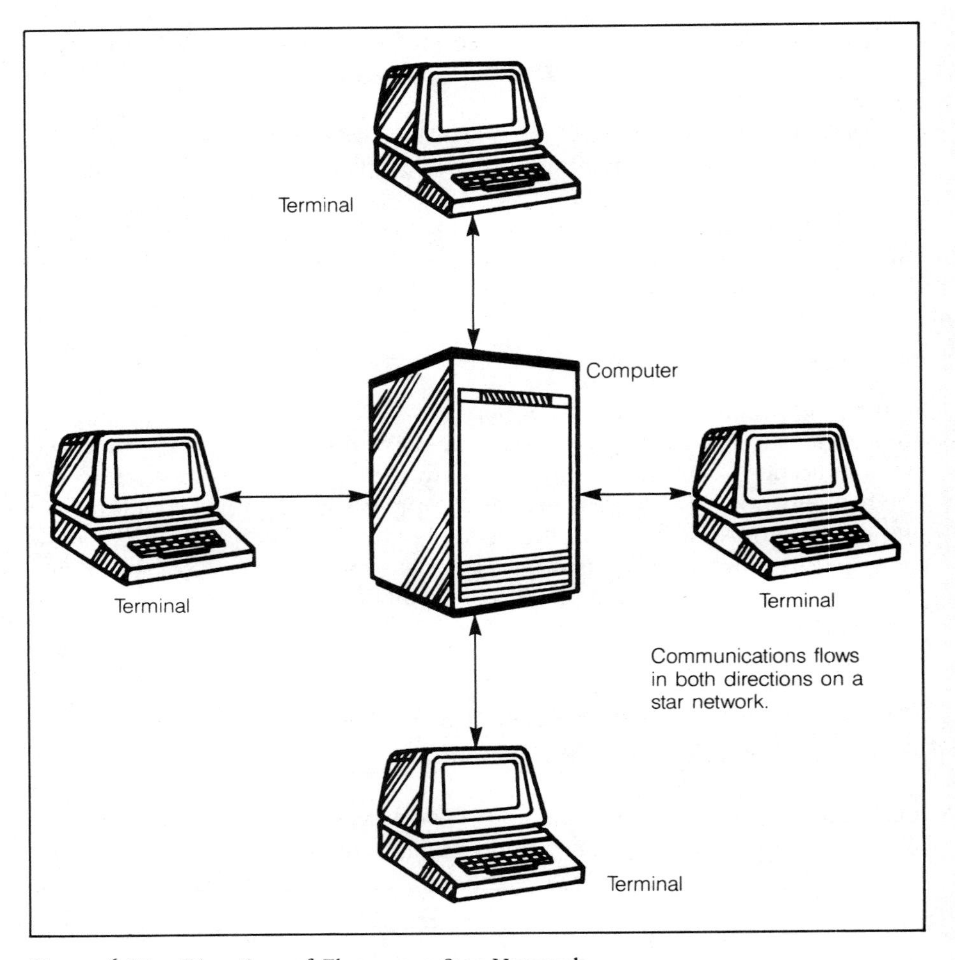

Figure 6-17. **Direction of Flow on a Star Network**

access techniques are required to keep the communications to and from each device separated and to be sure they do not interfere with each other. There are a number of different methods of handling this problem. Four widely used techniques—polling, token passing, slotting, and contention—are discussed in the following subsections.

Polling and Selecting

Polling is the process of asking the devices attached to the line if they have any messages to send. There are several different methods of polling, all of which are implemented in a "master-slave" environment. That is, there is always one master device (usually the central computer) that polls the slave or remote

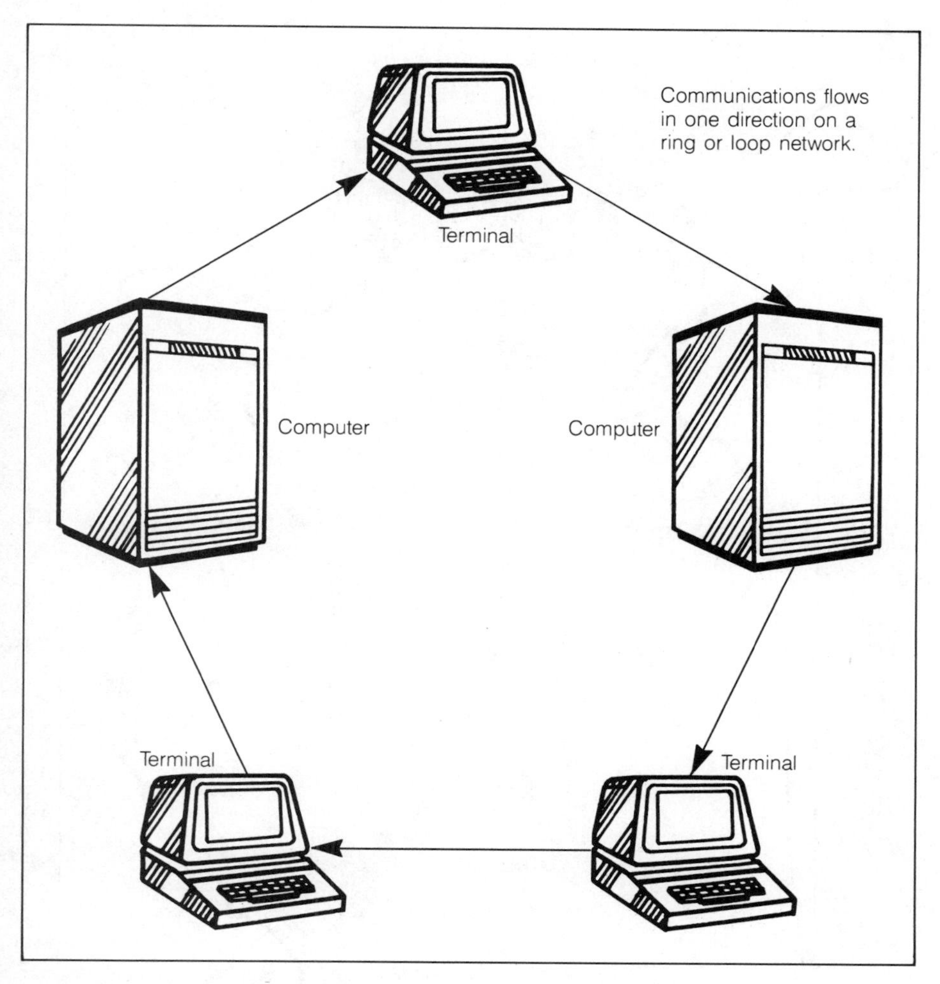

Figure 6-18. Direction of Flow on a Ring or Loop Network

devices (usually computer terminals) to see if the remote devices have any messages to send to the master.

The three most popular methods are roll-call, serial, and hub polling. We will describe these methods a little later in this chapter, but they all share certain characteristics. In all of them, each device on the line is assigned a unique address such as "01" or "AA." Before a device can send a message to the computer, the device must have received a special message or "poll" asking if it has any messages to send.

If a cluster controller or other station with multiple devices attached to it is used, the central computer polls the cluster controller rather than the terminals themselves. In this configuration, the computer makes a "general poll"—in other

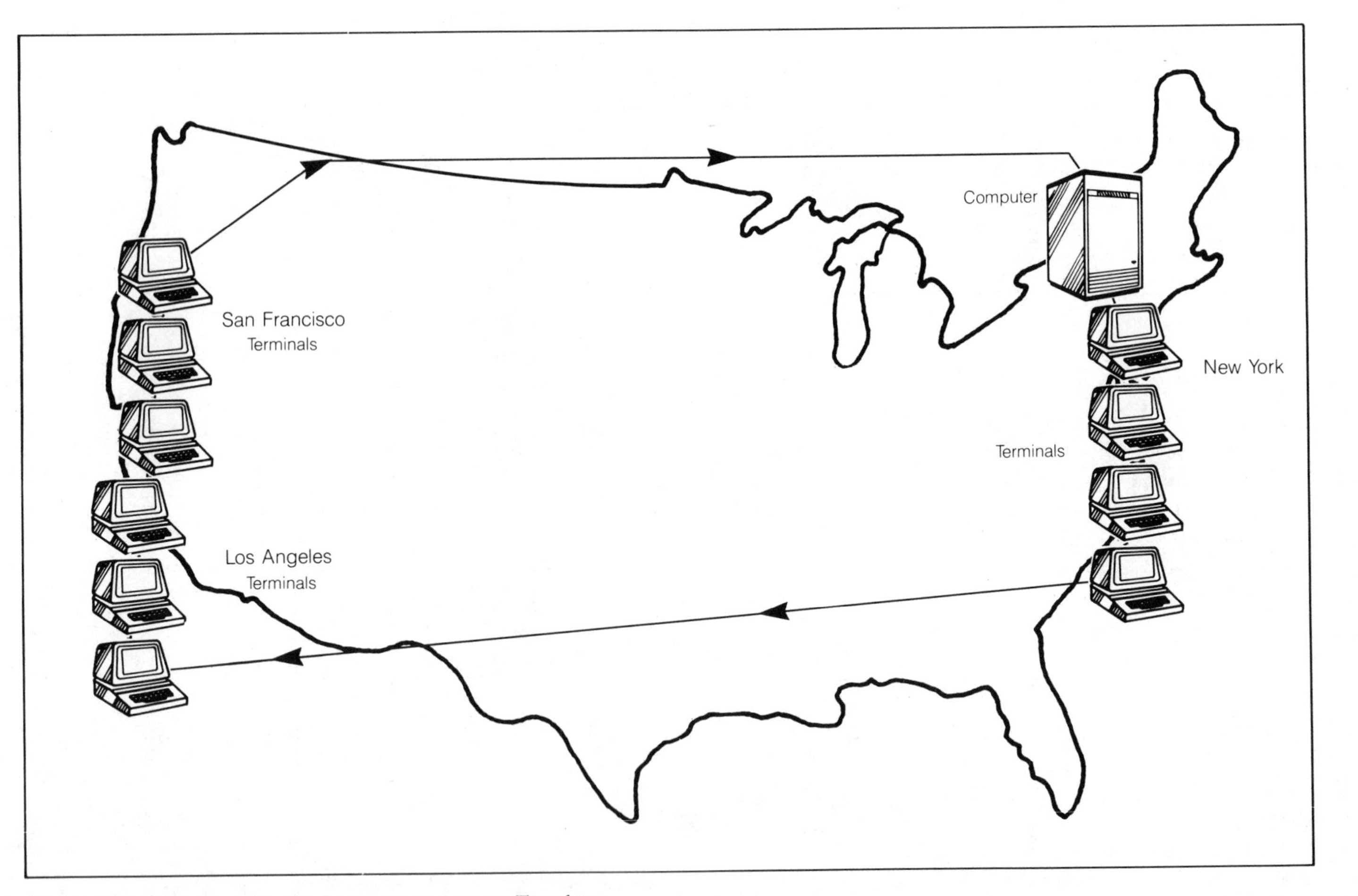

Figure 6-19. Three Cities Connected in a Loop Topology

words, a poll for messages from *any* of the devices attached to that particular station—or it can make a "specific poll"—that is, a poll for one specific device attached to the station.

What Is Selecting? When the computer wants to send a message to a device on the line (either a message transmitting data or a poll), the computer includes the address of the device in each message. The message is put on the line and "broadcast" to all of the devices on the line. All these devices "hear" each message, but only the device whose address is the same as the address in the message acts on or responds to the message. This technique is called *selecting* because a particular device is selected to receive the message (see Figure 6-20). In some systems, special addresses can be used to send one message to a group of devices.

Two Forms of Selecting There are two basic forms of selecting. One form is known as *select-hold* or *select-verify*. In this approach, the computer, in order to verify that the desired station is on the line and able to receive a data message, sends a special "are you ready?" message to the station (this is called "selecting the station"). If the station is able to receive a data message, the communications path to that station is maintained or "held," and the message or messages are then sent. This two-step approach is obviously time consuming and inefficient, although it is a rather simple method to ensure that the station will receive a data message that is sent.

The second approach is called *fast-select*. With fast-select, the computer sends the data message immediately and then uses complex error-recovery procedures if the message is not received.

The Problem of Overhead One problem with the polling and selecting network access method is the time delay and overhead involved in the polling process that reduces both the data-carrying capacity of the circuit and the computing capacity of the central control computer. Remember, a message must be sent from the central computer to each station or device on the line in order to inquire if that station or device has any messages to send.

The inquiry message itself puts a certain number of characters on the line; then the computer must wait for the station or device to answer, even if it has no information to send at that time. If the station is turned off or is malfunctioning and does not answer, the computer waits for a period of time known as a "time-out" period and then goes through error-recovery procedures, which usually entail sending more polling messages to see if the station will answer.

Additionally, if the circuit is long, say from New York to San Francisco or if it is a satellite circuit, the propagation delays (the time it takes for the message to get from the transmitting point on the circuit to the receiving point) can also eat up much of the data throughput capability of the circuit.

In spite of these problems, polling is a popular way to implement multidrop circuits. The following paragraphs look at the three versions of polling that are used.

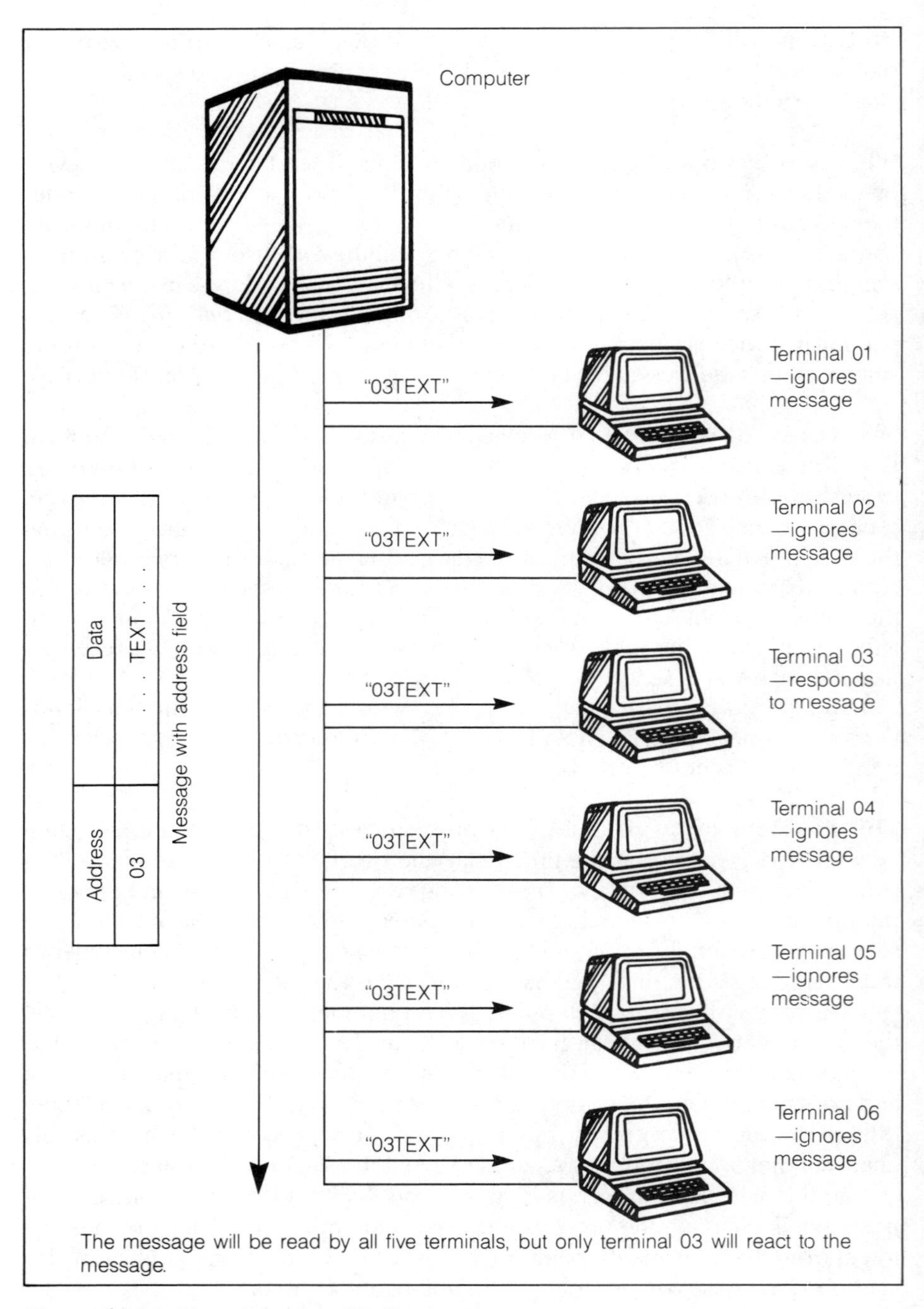

Figure 6-20. How Selecting Works

Roll-Call Polling With this method, the computer maintains a list or "roll" of the stations' or devices' addresses. The computer sends a poll to each station (which may be a single device such as a CRT terminal or a device such as a cluster controller with terminals or printers attached to it) on the line in the same order that the stations are on the roll.

The order in which the device addresses are listed in the roll does not have to be the same as the physical order of the devices on the line. In fact, some devices can be listed on the roll more than once, so they get more frequent attention from the computer than do other devices.

Serial Polling This is another form of polling that lowers the overhead of all the back-and-forth messages and shortens the propagation delays. Figure 6-21 illustrates serial polling.

When the central computer wants to inquire if any stations have any messages to send, it sends a polling message to the first station in the series (01 in this example). The message is received by the first station on the line, and, if it has data to send back, the data are sent. If not, either no message is sent or a "no data to send" message may be sent, depending upon the protocol being used. The first station then changes the address in the polling message to the address of the next active station in the series (02 in this example). Station 02 then repeats this process.

This technique reduces both the overhead and propagation delay problems associated with roll-call polling. Remember, however, that, in addition to changing the polling message, the stations must be able to keep track of the next active station in the series and know when a station is turned on or off; thus, the stations require a considerable amount of intelligence.

Hub Polling Hub polling is very similar to serial polling, but the network is configured with controlling stations or hub stations that then have remote stations connected to them, as shown in Figure 6-22. The central computer polls the hub, which in turn polls each of the remote stations associated with it, and then the hub station changes the address to that of the next hub and sends the poll message on. Alternatively, after one hub has finished polling, the central computer could poll the next hub.

Variations in Polling and Selecting The examples I've given you of roll-call, serial, and hub polling as well as of select-hold and fast-select demonstrate the basics of these concepts. Various manufacturers of data communications systems and equipment have implemented these concepts in ways that may involve slight twists or differences from the basics presented here. Nevertheless, with an understanding of the basic polling and selecting concepts, you should be able to understand the specific scheme used by any manufacturer.

Token Passing A second approach to network access is called *token passing.* Token passing is similar to serial polling, except that basic token passing is

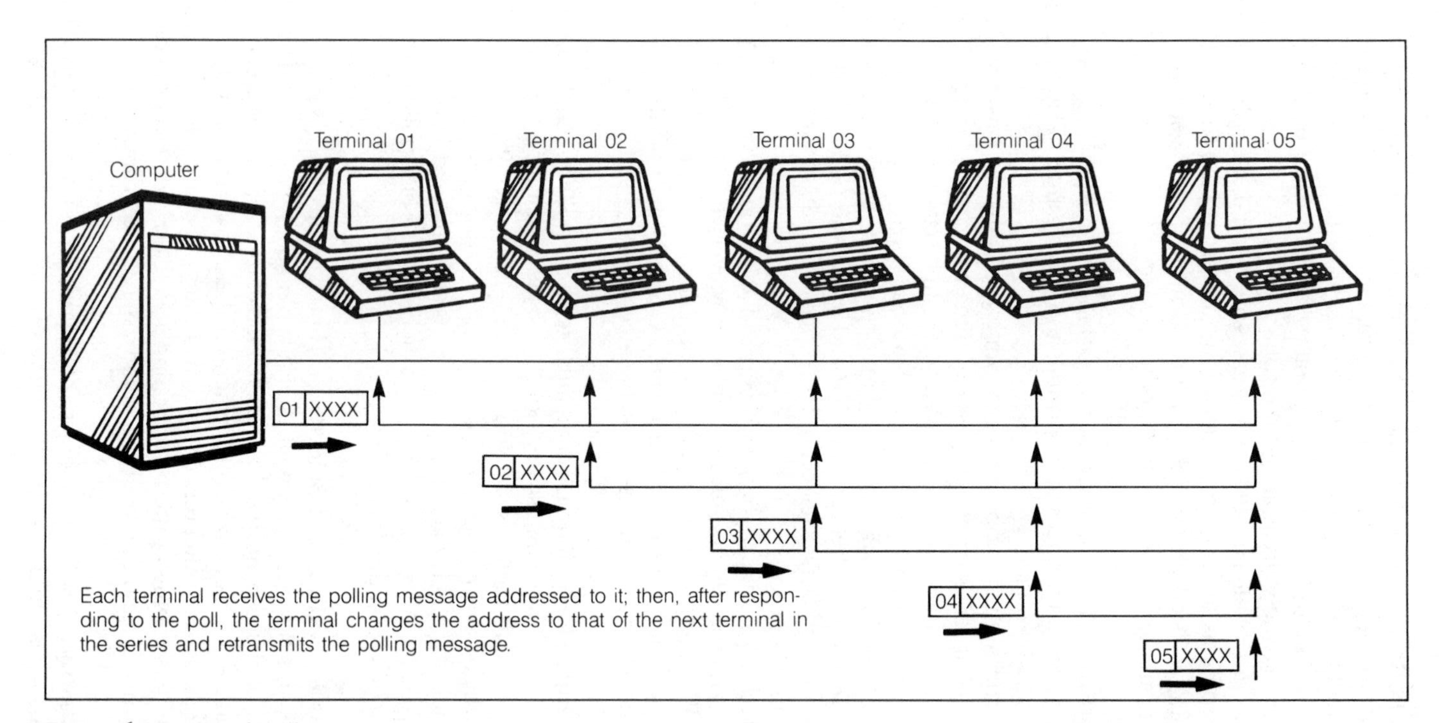

Figure 6-21. Serial Polling

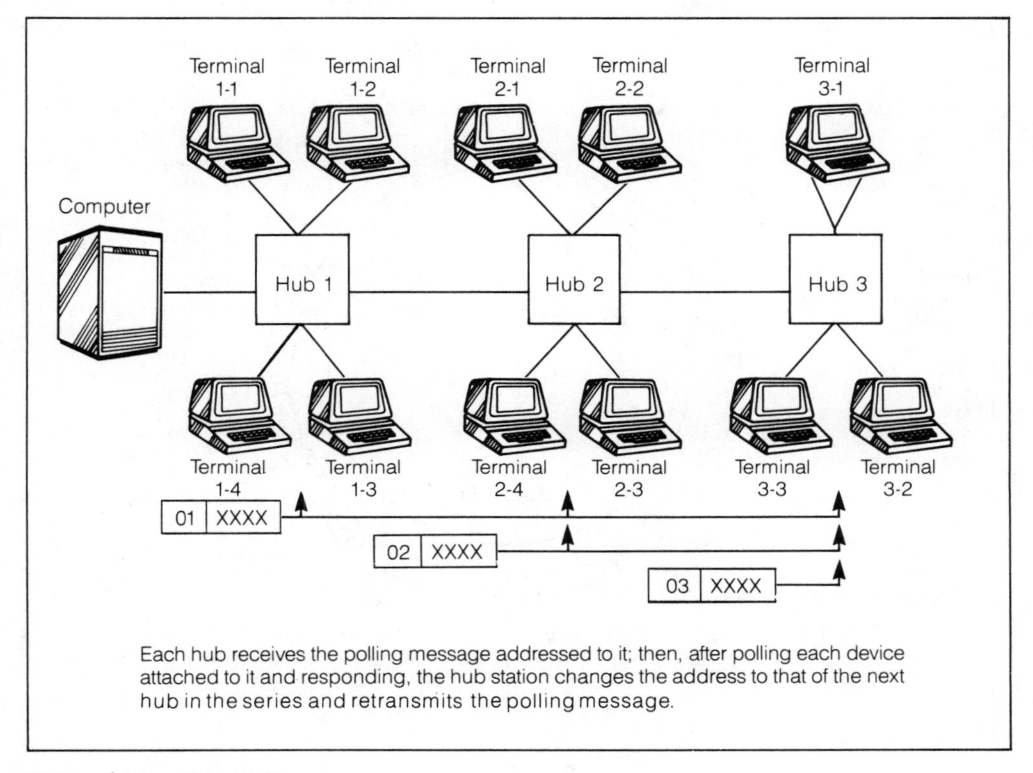

Figure 6-22. **Hub Polling**

implemented on ring or loop networks where only the next station on the line can "hear" a message.

Here's how token passing works: the central computer or controlling node in a network generates a special message called a *token,* which is nothing more than a unique pattern of bits (such as 1111111) that is defined as that system's *token.* The first station on the line reads all data that come down the line. If the data are not for it (that is, if they do not have the first station's address) or are not the token, then this station retransmits the data to the next station. Only when the station "sees" the token can it transmit its data. After the station transmits its data, it places the token back on the line so the next station on the line will "see" the token. When the next station receives the token, it transmits any data it may have to send and then inserts the token on the line, and so forth (see Figure 6-23).

With token passing (unlike serial polling), the network must be arranged so that the token message put on the line by one station does not automatically go to all stations on the line but only to the next station in sequence and that it then is regenerated by the station holding the token. Thus, basic token-passing schemes are implemented on a ring or loop architecture rather than on a bus architecture.

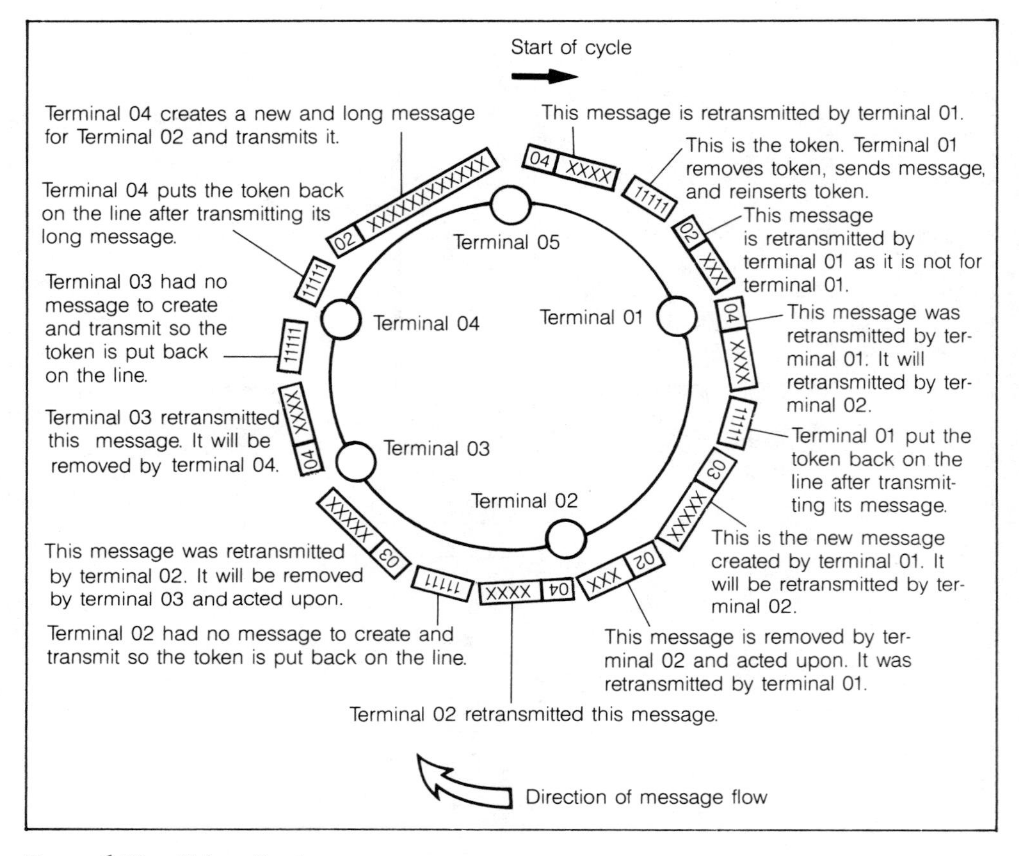

Figure 6-23. Token Passing

Remember that in serial polling, the polling message contains the address of the station being polled, and this address is changed by each successive station. Thus, serial polling can easily be implemented on bus networks since, although all stations hear the poll, they will all ignore it unless the address in the poll message is theirs. With basic token passing, the token is always the same bit pattern and thus must be transmitted to only one station at a time.

Token Passing on a Bus A form of serial polling implemented without a master/slave relationship (that is, any node can send a message to any other node rather than only being able to send a message to the control station) is also referred to as token passing. In this version of token passing, the token contains the address of the station that it is meant for as well as the special bit pattern that establishes it as the token. The receiving station (the station with the same address as that in the token) sends its message or messages and then changes the address in the token to the next terminal in the series and puts the token back on the line. With this "addressed token" approach, token passing can be implemented on bus networks or other topologies where all stations hear every message on the line.

The Patent Issue In 1981, the U.S. Patent Office issued a patent for a token-passing local network scheme to a Swede named Oloti Soderbolm (this patent references over twenty other token-ring patents). In 1980, IBM bought an unlimited worldwide license to use Soderbolm's patent in its products. Additionally, IBM has indicated that it has token-passing patents that it is willing to license, and AT&T's "Farmer-Newhall" patent may also restrict the available token-passing schemes. Soderbolm's patent, IBM's patents, AT&T's patent, as well as other token-ring patents may or may not be an impediment to the growth of token-passing schemes, depending upon the cost of licenses and the applicability of these and other patents to a particular vendor's token-passing scheme.

Make sure that your vendor is aware of the problems these patents may cause, and ask the vendor for "patent indemnification." If you are dealing with a small company with limited financial resources, be particularly careful. Finally, check the data communications literature for the latest developments in token-ring patents.

Slotted Network Access

With the *slotted network access* method, the transmission time is broken into equal-length slots or time periods. Each time period or slot starts with a flag or unique set of bits, which tells each node that the slot is either empty or full. When the node wants to send a message, it reads each slot's flag as the slot comes by and sees if the slot is full. If the slot is full and the message is not for that station, the station retransmits the message in the slot. If the message is for that station, the message is read, the station resets the flag to full, and it replaces the original message with the new message it wants to send. If, on the other hand, the flag indicates the slot is empty, the station can then set the flag to full and transmit its message.

Slotted network access, like token passing, is normally implemented on a ring or loop topology. With the token-passing method, once a station receives the token and is "in control" or able to transmit, the station can transmit a variable-length message; that is, the station can send a short message or a long message before it transmits the token, thus passing control to the next station.

In the slotted method, messages must all be of a predefined length—that is, the length of the "slot." If the station is sending a shorter message, the message can be "padded" with blanks or null characters or an end-of-message symbol or control character can be used. If the message is longer than the slot, the message must be segmented into shorter pieces and then put back together at its destination. This, of course, requires sophisticated network control equipment (see Figure 6-24).

Contention

With the *contention network access method,* all peripheral devices try to access the line to the computer whenever they have anything to send, rather than wait for a token or a "do you have anything to send?" polling message. If two or more devices try to send messages at the same time or close enough so the messages overlap, the messages will interfere with each other, and the computer will not understand the message. Thus, some technique needs to be used to avoid message overlap.

Carrier Sense Multiple Access (CSMA) The most popular technique used in contention systems is known as *carrier sense multiple access (CSMA).* With CSMA, when a node wants to send a message, it listens on the line for a tone or "carrier." If there is no carrier, then the line is presumed to be clear or available. The node then puts a carrier signal on the line and sends its message. After the message is sent, the node turns off the carrier.

Collisions with CSMA One of the problems with the CSMA system is that collisions can occur due to propagation delay. To understand this, let's look at an example (see Figure 6-25). If terminal 3 sends a message to the computer, there will be a short delay before the carrier and message reach terminal 1 on their way to the computer. If terminal 1 listens, hears no carrier, and then sends a message while the carrier and message from terminal 3 are on their way down the bus, the message from terminal 3 will collide with and interfere with the message from terminal 1.

CSMA with Collision Detection A variation of the CSMA technique designed to overcome this problem is called carrier sense multiple access with collision detection (CSMA/CD). CSMA/CD does not stop the collision but rather detects it and then retransmits the message. CSMA/CD is the network access technique used with Xerox Corporation's local area network (LAN) concept called Ethernet. (Ethernet will be discussed in detail in a later chapter.)

POINTS TO REMEMBER

This chapter has provided you with a basic understanding of the current most commonly used network topologies and access methods. We looked at five topologies: the star, mesh, bus, tree, and ring or loop. We then looked at four

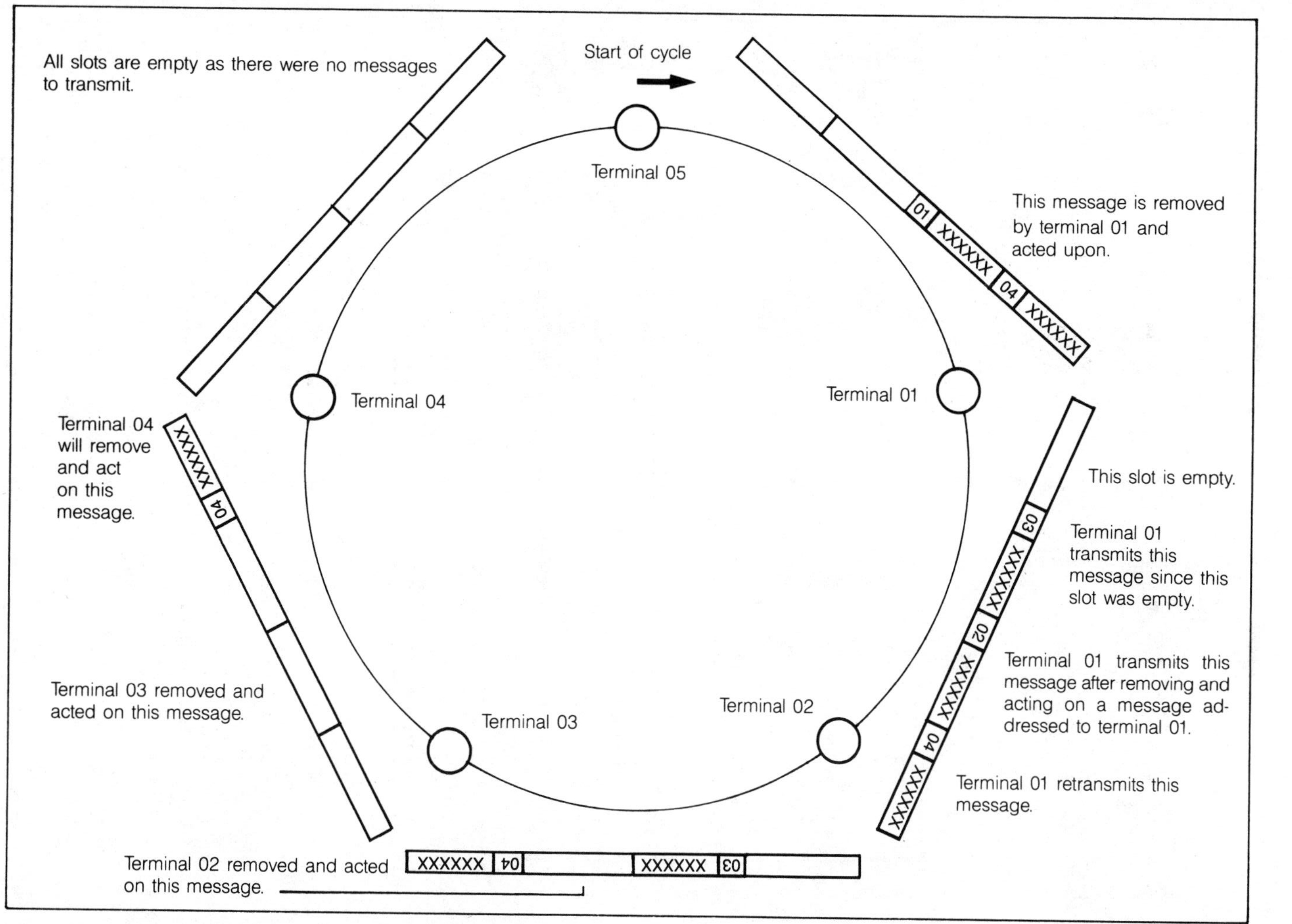

Figure 6-24. The Slotted Network Access Method

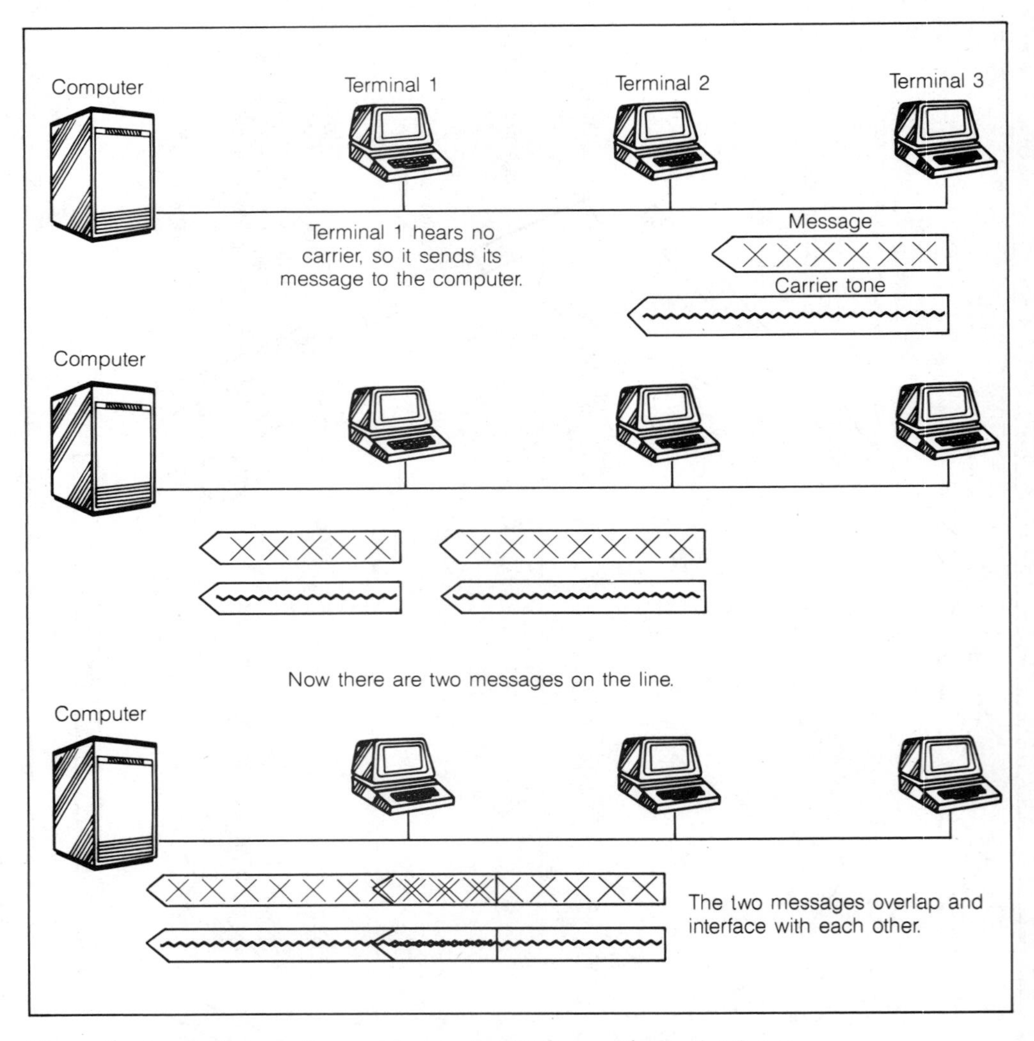

Figure 6-25. Collisions Due to Propagation Delays with Carrier Sense Multiple Access

network access methods: polling and selecting, token passing, slotting, and contention.

Manufacturers of data processing and data communications equipment often require specific topologies and access methods to be used with their equipment. You will often not have a choice (or at least very much of a choice) in selecting the topology and access method you want if you have already selected a particular vendor for your data communications equipment. Thus, you need to understand the choices available and to be able to evaluate how these choices fit into your individual data communications needs before you select a vendor.

The topologies and their related access methods differ in their cost for network control equipment as well as in their cost for communications circuits. They will also differ in terms of such other parameters as data rates, reliability, the ease with which a bad or broken circuit can be bypassed, and so on. Be sure to discuss the implications of the network that equipment vendors recommend (or require) before locking yourself into a specific family of data processing or data communications equipment.

The next chapter gives you more information about networks to ensure that you are well informed before you go to the vendors themselves.

7

NETWORKS—OPTIONS FOR ORGANIZATION

In addition to considering the topology and access method to be used on a network, you must also make important decisions about how the network is organized. You must choose, for example, what kind of communications line to use. Will you use dedicated ("leased") circuits, or will you use dial-up or switched circuits? What do you do if one of your dedicated circuits stops working and it is important for the remote device or devices attached to that circuit to gain access to the network? How can you prepare, in your network design, for the possibility that the load to a particular computer, device, or group of devices will increase beyond normal limits? What do you do when users need maximum flexibility in reaching many computers or data bases? This chapter provides answers to these questions and gives you a sense of your options for network organization.

CHOOSING YOUR COMMUNICATIONS LINES

One of the first organizational issues you have to address is the decision to use dedicated or switched circuits in your network.

Look at the star network at the top of Figure 7-1. Note that each terminal is connected to the computer by its own communications circuit. If, on the other hand, the terminal is not being used very often, then you can connect it to the computer with a *switched circuit,* using a dial-up telephone line and getting data speeds of up to 4800 bps, just like you would dial up and connect your telephone instrument to someone else's for an ordinary telephone conversation (see the bottom of Figure 7-1).

The Advantages of Switched Circuits

The main advantage of switched circuits is that you are only charged for the time you actually use them. Thus, they are much less expensive than dedicated lines if you only need to use them infrequently.

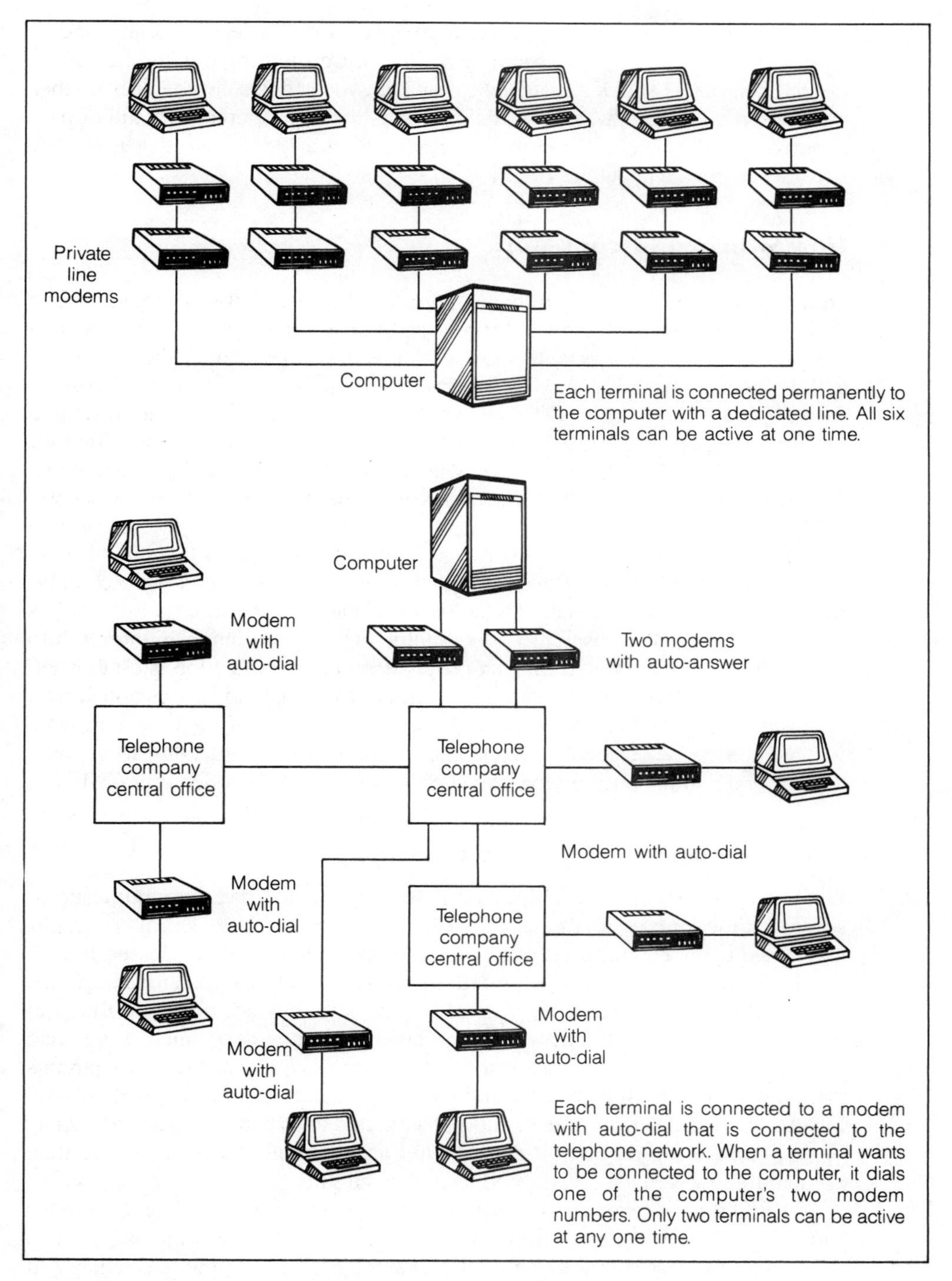

Figure 7-1. Two Ways to Connect Six Terminals to a Remote Computer

They also allow you to connect a peripheral device easily to any one of several different computers by simply dialing or transmitting the number of the desired computer. And, if a switched circuit fails, you can usually establish another connection quite quickly. Finally, dial-up voice lines can operate with full-duplex protocols at 4800 bits per second (or less) and at an error rate that is usually considered very acceptable.

How Switched Circuits Work

Dial-Up Voice Lines To connect a peripheral device to the computer over a dial-up voice line, you establish the connection using a telephone in the normal manner. After you have established the connection (that is, after the telephone at the other end has rung and been answered), you throw a switch that bypasses the telephone and connects the device directly to the telephone line through a modem. The party at the other end also throws a switch connecting the computer to the same line through its modem. With some systems, the dialing, answering, and switching of the data communications equipment onto the line are all accomplished automatically.

Dial-Up Digital Lines If dial-up digital lines are used instead of voice lines, the procedure is very similar. The address of the device you are contacting is typed on a terminal, transmitted over the line by your computer or entered into a signaling device provided for that purpose, and the connection is established directly to the data communications equipment rather than first establishing a voice connection. Of course, on a digital circuit, modems are not used; instead, you use some type of signal conversion unit, sometimes called a line access unit, digital service unit, data service unit (DSU), or a channel service unit (CSU).

The Advantages of Dedicated Circuits

Dedicated circuits are available from many sources for data communications; some of these sources are discussed in Chapters 9 and 11. If the dedicated circuits are leased from a common carrier, they are often referred to as "leased lines."

In many applications, a dedicated line is a better choice than a dial-up line. Dial-up lines may be routed differently every time they are used and thus can have different electrical characteristics each time; they cost much more than leased lines if they are used many hours during the day; and dial-up voice circuits can only be used with half-duplex protocols if the data speed is faster than 4800 bits per second. Also, the special modems needed to transmit data at 1200, 2400, or 4800 bps in full duplex on dial-up lines are generally more expensive than comparable modems for dedicated four-wire circuits.

In addition, it takes time to set up the connection with dial-up lines. This can be a real problem for systems with very short but frequent messages (such as the systems that run automatic tellers) since the set-up time, particularly in the standard telephone voice network, can be longer than the transmission time for the message. However, there are some networks that have been designed especially for data communications that can switch circuits electronically in a fraction of a second, and thus set-up time is not as much of a problem as on the voice telephone network. These systems are referred to as fast-connect networks or fast circuit-switching networks.

Dedicated circuits require no connection set-up time, are not frequently rerouted by the common carrier, have a fixed cost no matter how many hours they are used, can support full-duplex protocols at speeds of up to 16,000 bps, and generally use less expensive modems.

How Dedicated Circuits Work

Four-Wire Circuits Dedicated circuits are available as four-wire circuits with one pair of wires for transmitting and the other for receiving, they allow full-duplex transmission at all speeds that the line can handle for only a small surcharge over the cost of two-wire circuits. These four-wire dedicated lines can use simpler and less expensive full-duplex modems than those required for full-duplex transmission on two-wire or dial-up lines.

On the other hand, an often-overlooked advantage of four-wire circuits is that one four-wire circuit can be converted easily into two full-duplex circuits by using special modems designed to transmit full duplex over two-wire lines. These special modems are available at 300 bps, 1200 bps, 2400 bps, and 4800 bps. Although these special modems cost more than ordinary ones, they can be very cost effective when used with an expensive dedicated four-wire circuit such as one from San Francisco to New York. In fact, they can often pay for themselves in only one or two months.

In order to derive two full-duplex circuits from one four-wire circuit, you must be able to connect the transmitters from both modems to the same two-wire transmit pair and both the receivers to the same two-wire receive pair (see Figure 7-2). You also must be able to set the frequencies used by the modems so that one pair of modems (one at each end of the circuit) will transmit on the high-frequency band and receive on the low-frequency band while the other pair of modems will transmit on the low-frequency band and receive on the high-frequency band. If you are going to use two two-wire full-duplex modems at each end of the circuit, be sure the manufacturer has made provisions both for changing the frequencies and for strapping the receive and transmit sections of the two modems together.

Multipoint Circuits Dedicated lines are almost always used on multipoint or multidrop circuits. One of the primary reasons for using multipoint circuits is that one line or circuit can serve many terminals. The result of many terminals using this one line is to increase the amount of data traffic on the line above what would be possible on a point-to-point line with just one of the terminals

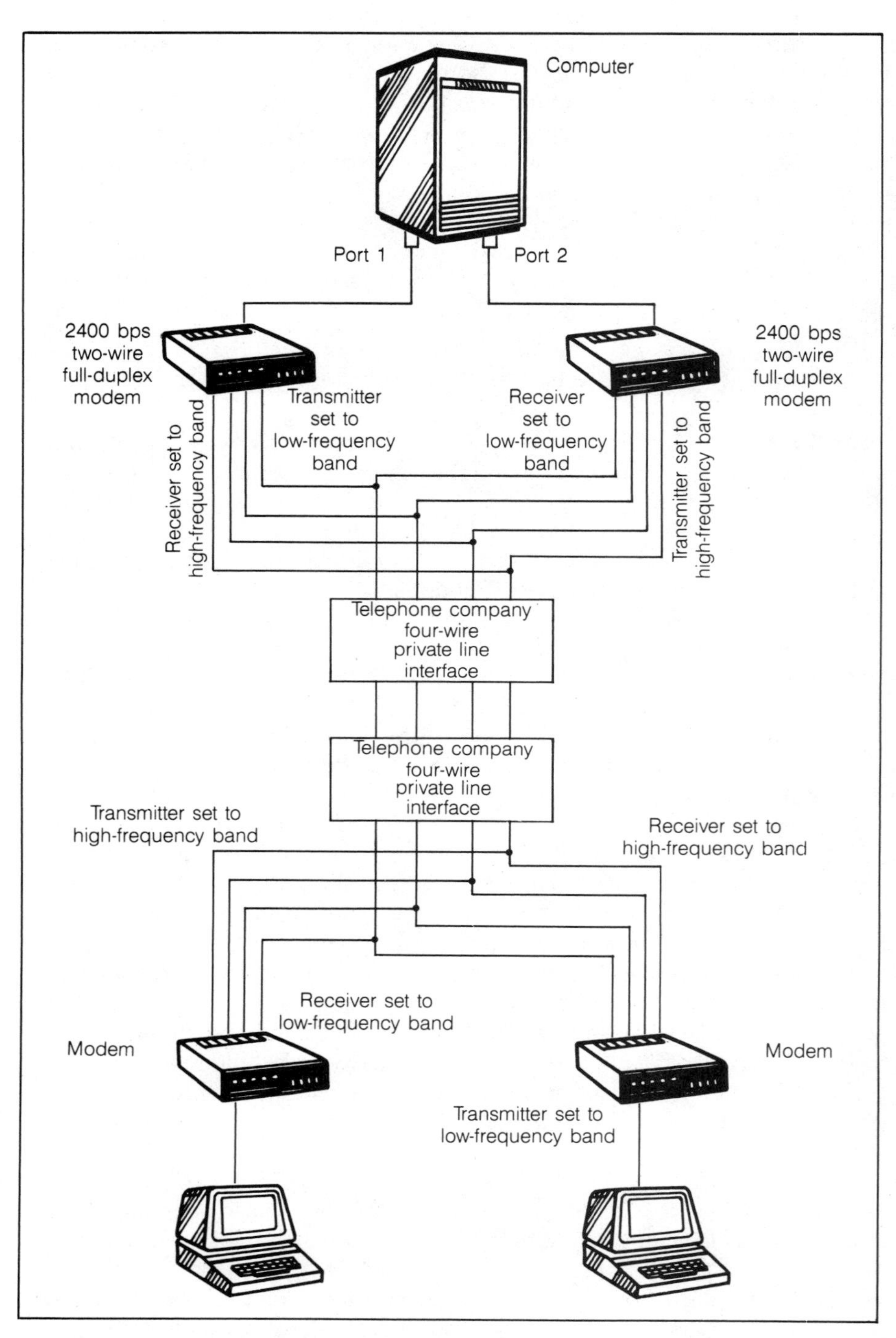

Figure 7-2. Using Four Special Two-Wire Modems to Obtain Two Full-Duplex Circuits from a Single Four-Wire Circuit

attached. Therefore, multipoint lines tend to be more heavily used than point-to-point circuits.

Multipoint circuits also tend to be in use over relatively long periods of the day. This heavy usage alone makes dedicated circuits more economical than dial-up circuits for multipoint networks.

How to Use Dial-Up Lines for Multipoint Communications

AT&T historically has not wanted to provide the bridging capability at their central offices that would allow dial-up lines to form a multipoint network. However, the bridging can take place at your own facilities, and it may be very practical to have one or more legs of a multipoint circuit consist of dial-up lines, especially if those legs are used infrequently or if the dial-up circuits are part of a private company network that can be used during off hours for no additional charge (see Figure 7-3). In Figure 7-3, a terminal in San Diego is connected to the dial-up telephone network by a two-wire dial-up line modem. The connection is made through the telephone company office just as with a voice telephone call. At the receiving end, the telephone is connected to a two-wire dial-up modem that converts the signal back into digital form. The digital signal is then converted back to an analog signal and connected to the four-wire multidrop circuit by means of a four-wire leased-line modem. For example, such companies as Westinghouse, IBM, and General Electric have private networks that usually consist of leased lines from the telephone company hooked up to form a *private* dial-up telephone network. In the evenings, when these telephone networks are not needed for voice traffic, the lines can be used for data communications.

Dial-up lines are also frequently used as part of a multipoint circuit when the dial-up lines are used to back up or substitute for a section of the dedicated multipoint line that is out of service.

PROVIDING BACKUP IN CASE A LINE FAILS

If a dedicated data circuit fails, it can be a very serious problem for the user. This section looks at various approaches you can use for backup and redundancy so that data communications can be maintained in the event of a circuit failure.

Dial-Up Backup for Dedicated Lines

When a dedicated or leased line becomes unusable (either because it becomes too "noisy" or because it begins to fail intermittently or shuts down totally), it may be hours or even days before the circuit is restored. For critical computer applications, it may be unacceptable for a remote terminal or other peripheral device to be unusable for that period of time. If your application falls in this category, you may want to consider using dial-up circuits as backup for your dedicated lines.

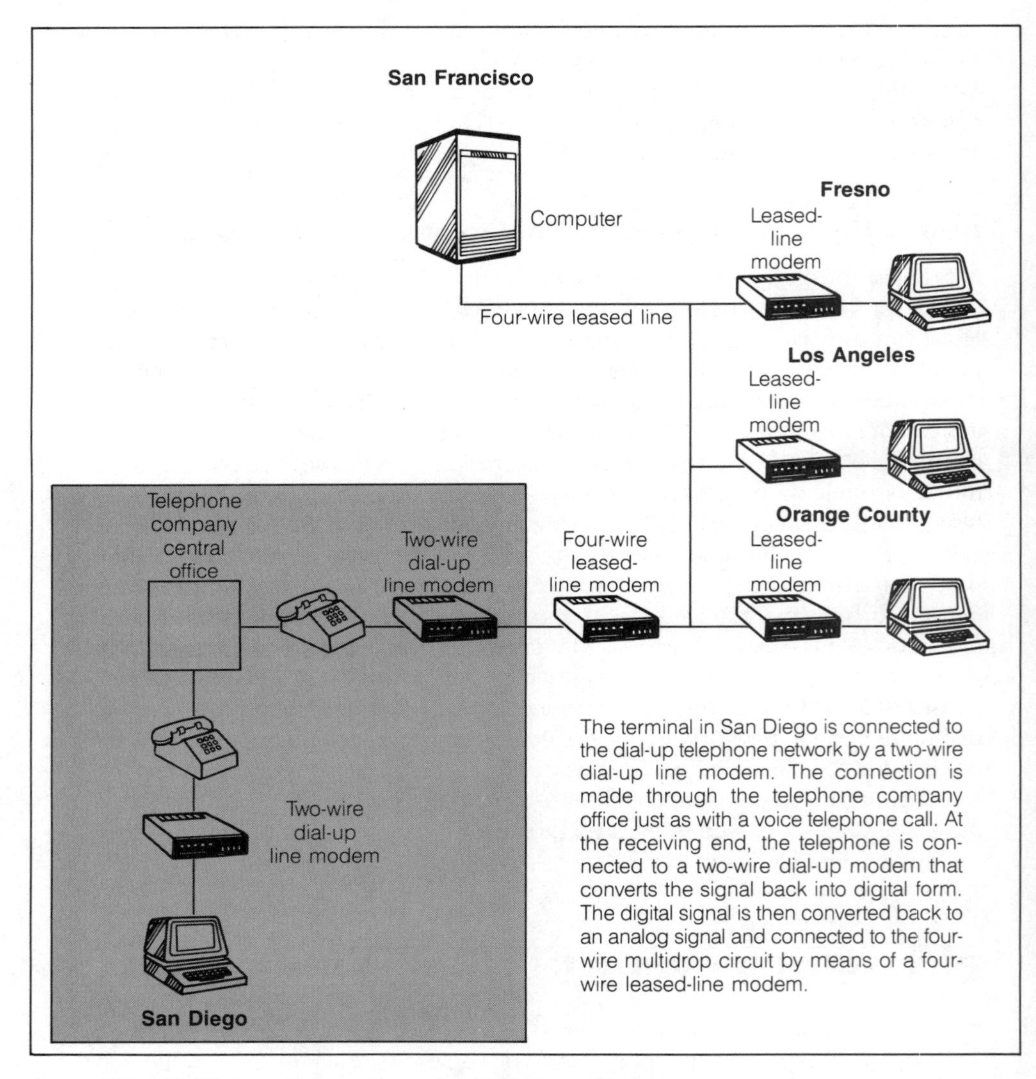

Figure 7-3. A Dial-up Line as Part of a Multidrop Circuit

Switching Modems If you are using dedicated point-to-point circuits with a half-duplex protocol and a data speed of 4800 bps or less, then using dial-up lines for backup will be straightforward except that you may have to change modems at both ends of the circuit before connecting the dial-up lines. Many half-duplex modems can work on dial-up voice as well as on dedicated voice lines. In fact, some modems can automatically establish the dial-up backup connection when the dedicated line fails. You should check with your supplier about this point before you buy modems if you are planning to use dial-up circuits as backup.

Backing Up a Four-Wire Full-Duplex Circuit If you are using four-wire dedicated circuits with a full-duplex protocol, you will run into the problem that dial-up circuits are two wire and can only support a half-duplex protocol except with special modems such as AT&T's 212A, Racal-Vadic's series 3400 (1200 bps) and model VA4400 (2400 bps), or Anderson Jacobson's 4800 bps modem.

There are at least four approaches you can use to overcome this compatibility problem; they are illustrated in Figure 7-4 and described in the paragraphs that follow.

1. The most popular approach is to use two dial-up lines, one for transmit and one for receive. Although using two dial-up lines is more expensive than using one dial-up line, this approach can provide full-duplex transmission at speeds of up to 9600 bits per second.

If you select your modems for this purpose to start with, you will not have to change them when you go to the dial-up lines as backup. Also, some modems can automatically establish the two dial-up backup connections needed when the dedicated circuit fails.

Some manufacturers, such as Racal-Vadic, supply modem-adapter and automatic-dialing units that allow you to convert a standard leased-line modem to one that can access the public dial network for backup. Because of the ease of using two dial-up lines for backup and because this approach can support full-duplex transmission at speeds of up to 9600 bps, two dial-up lines are the most commonly used backup method in this situation.

2. If the circuit speed of the failed circuit is 4800 bps or less, then you can substitute special modems that operate at those speeds in a full-duplex mode over two wires in order to use dial-up lines as backup. If the circuit speed is faster than 4800 bps, you could reduce the circuit speed, but this could require a change to your data link control (DLC) as well as to your modems.

3. You can convert the application to a half-duplex circuit; however, this approach will most likely slow your data throughput and may require a change in modems. In most cases this approach will also require a change to your DLC.

4. You can use a full-duplex digital dial-up circuit if this service is available between the two points served by your failed dedicated circuit. Digital circuits run full duplex and can handle speeds faster than 4800 bps.

Problems with Bridging a Dial-Up Backup Circuit If you are using a multipoint circuit, you face the problem of *bridging* (or connecting) the dial-up circuit to the multipoint circuit. Historically, AT&T has not wanted to provide this bridging service at its central offices. Other common carriers have occasionally provided the bridging at their facilities, and you may find such arrangements more available in the future. In the meantime, you can bridge the lines at your own facilities.

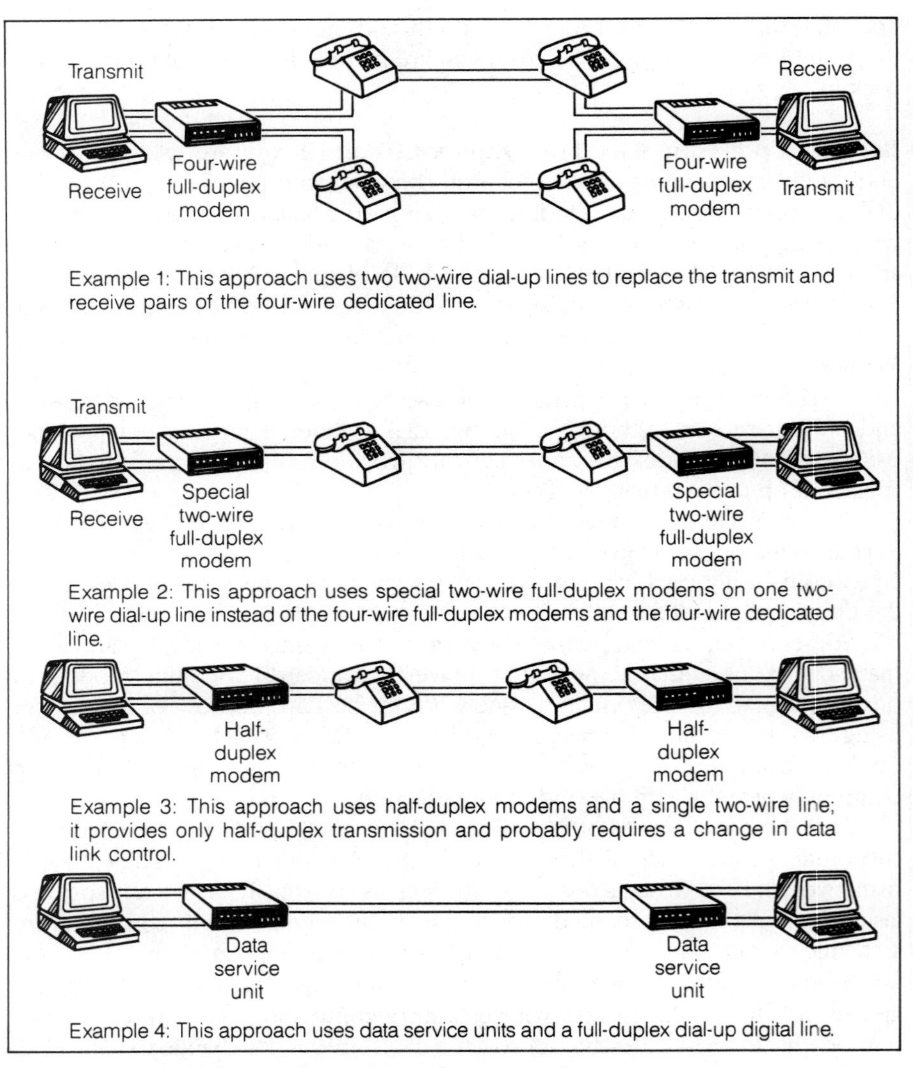

Figure 7-4. Four Approaches to Using Dial-Up Lines as Backup for a Four-Wire Dedicated Circuit

Special bridging equipment is available for use with circuits that are all analog, and different bridging equipment is available for use with circuits that are all digital.

If, on the other hand, the dial-up backup line is digital and the dedicated line is analog or vice versa, you will have to add a modem to convert the digital signal to analog (or vice versa) before bridging the two circuits (see Figure 7-5).

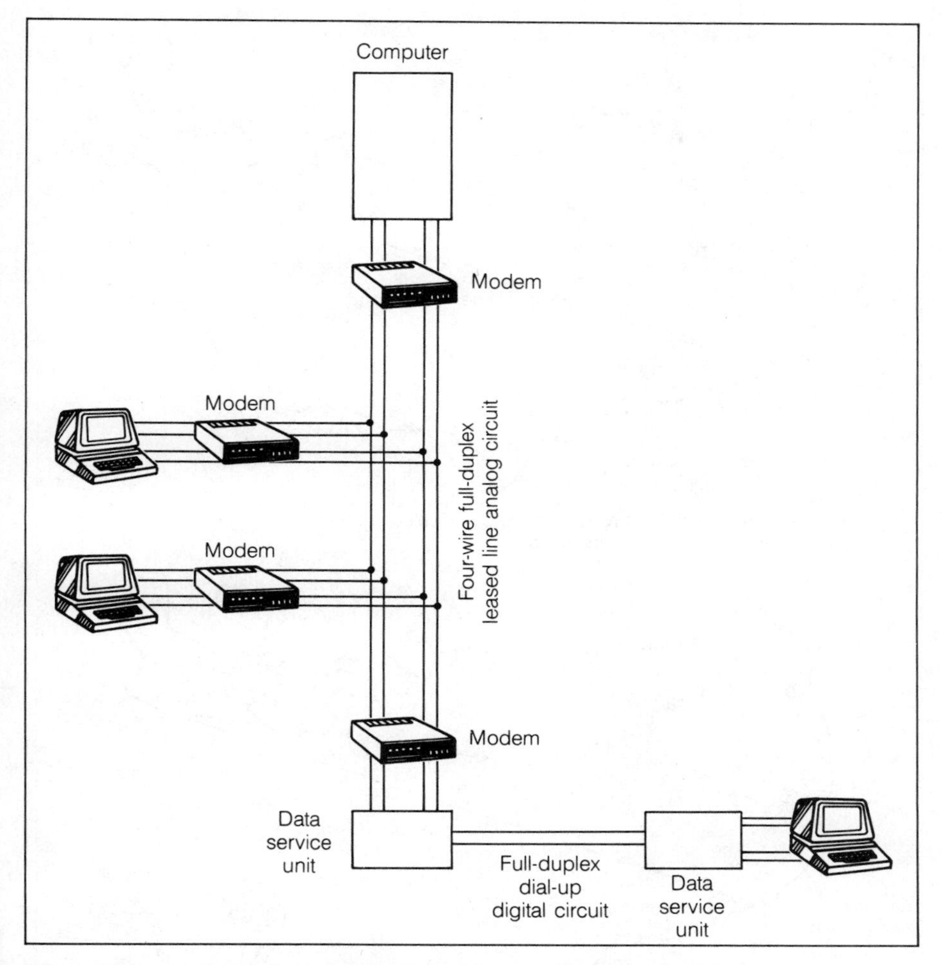

Figure 7-5. Bridging a Digital Backup Line to an Analog Network

Redundancy

Another approach to backing up dedicated circuits when they fail is *redundancy*. This means having multiple dedicated lines so that, if one fails, data communications can still be conducted over one or more of the other dedicated circuits. There are several ways to implement redundancy; I discuss four of them here.

Redundant Standby Link The first approach is to duplicate the line and possibly its associated equipment on any critical links (see Figure 7-6). If the normally used link fails, the redundant link is switched in. In Figure 7-6, the New

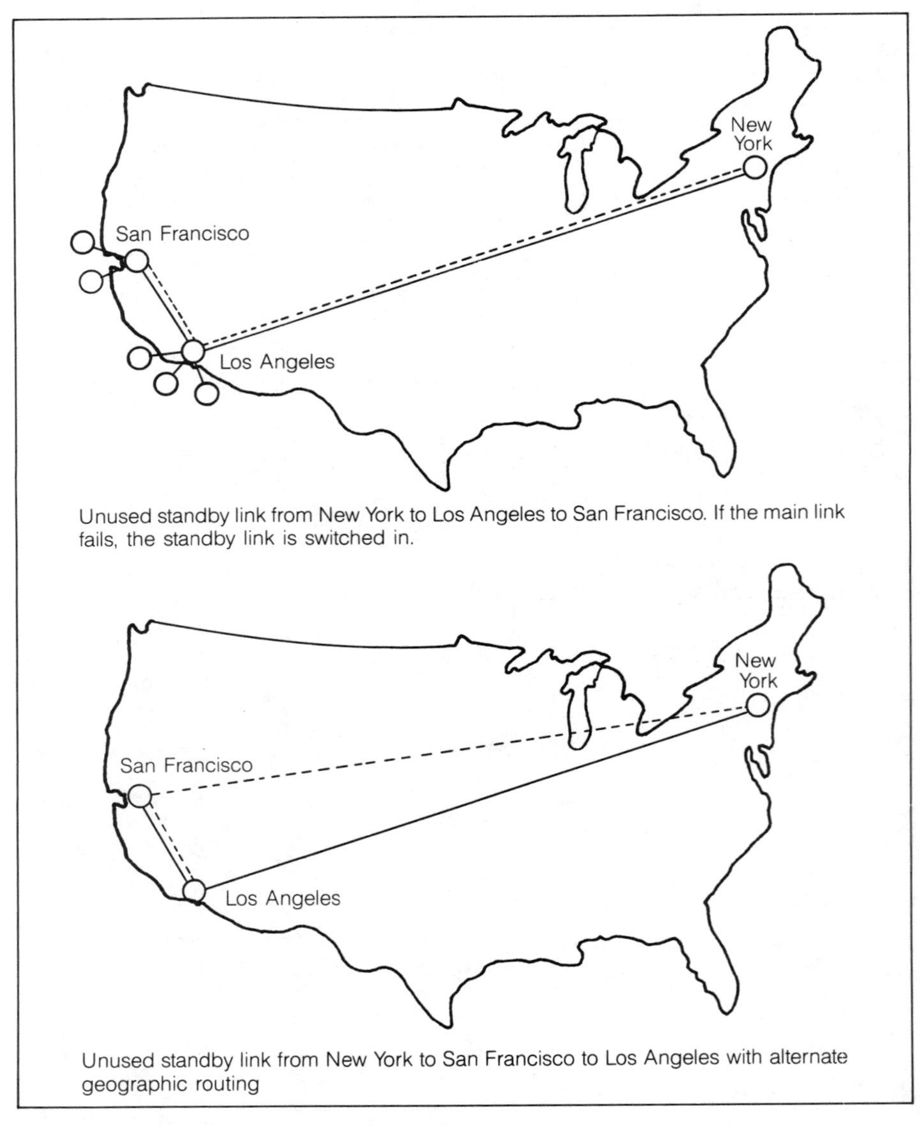

Figure 7-6. A Redundant Standby Link

York to Los Angeles link is backed up by a standby redundant link, since, if the N.Y.-L.A. link fails, all terminals in both Los Angeles and San Francisco would be inoperative. You could also have the redundant link between New York and San Francisco; this would give the additional protection of an alternate geographical route.

Redundant Link That Is Used Having an extra unused link is, of course, an expensive solution, so a second approach (really just a variation of the first) is to actually use the back-up line during normal operations.

In this approach, the communications load in normal use is split between the primary line and the back-up line by assigning certain terminals to each line. If either line fails, all terminals are switched to the remaining line. This approach does not reduce the costs but does provide superior throughput in the normal operational mode.

You should realize that, if one line fails, the throughput on the remaining line will still be at "standard" levels, since the network is designed to work without the redundant line. The normal throughput with both lines working is well above "standard."

Another advantage of this approach is that the network will have enough extra capacity to absorb unusually high levels of traffic if they occur. Both of these advantages are in contrast to a network that is designed to work on multiple lines at "standard" throughput levels and that experiences reduced throughput if one line fails (this approach is discussed under "Multiple Links").

Finally, using the redundant link also has the advantage over the standby-link approach of continually testing the backup link.

Multiple Links The third approach is to route the lines so that more than one line services critical facilities (see Figure 7-7) and so those multiple lines are all used during normal use to provide a "standard" level of service. If one line fails, the stations on the other lines in that facility will still work. For example, in Figure 7-7, if the line between terminal A in San Francisco and terminal B in Los Angeles fails, terminals C, D, and E can still be used at standard throughput levels.

A variation of this approach is to transfer the stations from the out-of-service line to those still in service. This provides service to all stations at the facility (in our example, terminals A and B would still be operative) but at the expense of reduced throughput to *all* the stations on that line.

Redundant Links Versus Multiple Links Note that the only difference between approaches two (redundant links used) and three (multiple links with "standard" service levels) is that approach two uses more lines than are necessary to provide the desired level of service during normal use but still provides the desired level of service during a line outage, while approach three provides a lower than "standard" level of service during a line outage.

Intelligent Communications Networks A fourth and very important approach to redundancy involves an alternate routing capability and is used in complex networks that employ advanced network control equipment. These networks are often referred to as intelligent communications networks (Figure 7-8). In the Figure 7-8 network, logical channels are created where no direct physical con-

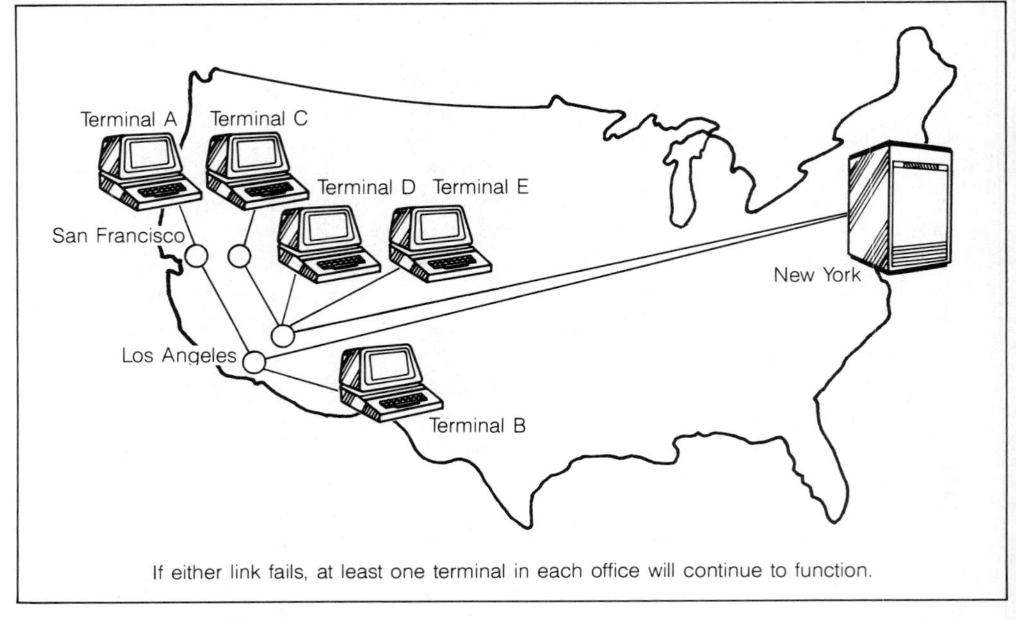

Figure 7-7. **Multiple Links**

nections exist, such as from Los Angeles to London. Also, alternate routes are available if a primary route is busy or if it should fail. For example, data can be sent from San Francisco to Chicago by routing it from San Francisco to Los Angeles to Kansas City to Chicago if necessary. IBM's Systems Network Architecture or SNA, DEC's DECNET, GT&E's Telenet (a public packet-switched network), and ARPANET (developed by the U.S. Department of Defense) are all examples of such intelligent communications networks.

In the network shown in Figure 7-8, there are computerized nodes located throughout the United States and Europe. Some computerized nodes have direct physical links to certain other nodes. Los Angeles, for example, has two links to Kansas City, two to San Francisco, and two to Miami, but no direct physical links to any other node on the system. (This topology, remember, is called a partially connected mesh.)

With intelligent communications networks (such as the SNA or DECNET network management systems), "logical channels" can be established where no direct physical connection exists by using the data communications hardware and software in the various computerized nodes. Thus, a message can be sent from Los Angeles to London by creating a logical channel from Los Angeles to Kansas City to New York to London or from Los Angeles to San Francisco to Chicago to Pittsburgh to New York to London and so forth.

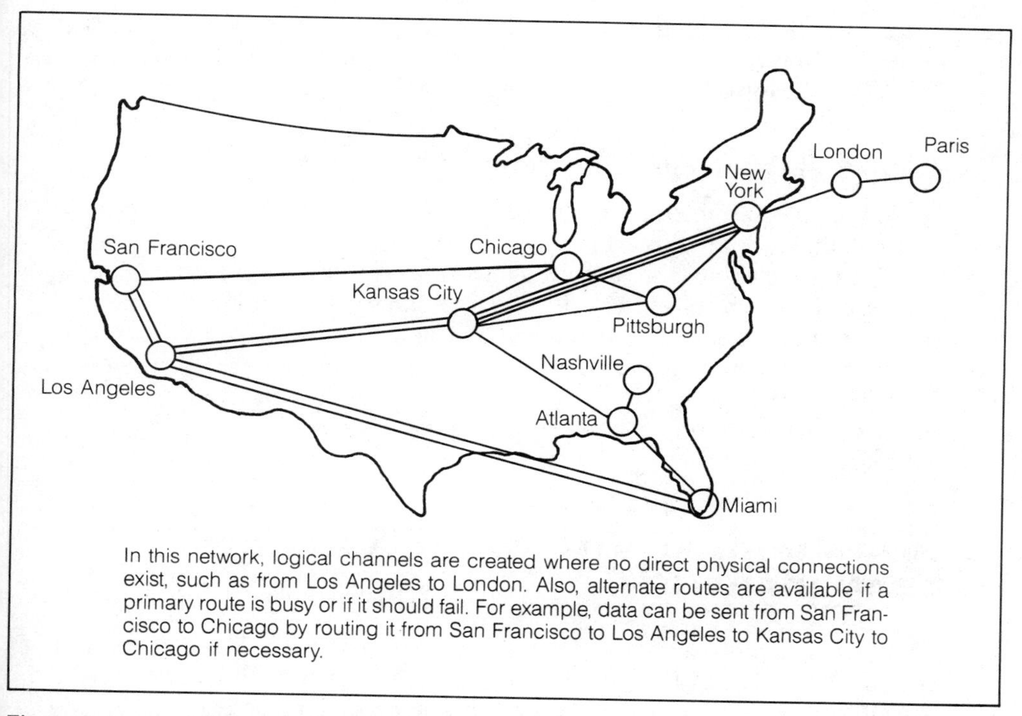

In this network, logical channels are created where no direct physical connections exist, such as from Los Angeles to London. Also, alternate routes are available if a primary route is busy or if it should fail. For example, data can be sent from San Francisco to Chicago by routing it from San Francisco to Los Angeles to Kansas City to Chicago if necessary.

Figure 7-8. An Intelligent Communications Network

Logical channels are the result of the hardware, software, circuits, and computerized nodes that allow data to move from one point on the network to another even if no direct physical circuit exists between those two points. (In the literature on packet-switching networks, you may find that logical channels are often referred to as "virtual circuits," "virtual channels," or "virtual connections.")

Look again at Figure 7-8. Note that if the link from Kansas City to Pittsburgh fails, the network management system could route the message from Kansas City to New York to Pittsburgh or from Kansas City to Chicago to Pittsburgh. Thus, *alternate routes* are a form of redundancy available with intelligent communications networks.

In a large complex network, two nodes are often connected by more than one direct physical link. Between Kansas City and New York, for example, there are three links shown in Figure 7-8. The network management system selects one of those three links when establishing a logical channel involving the circuit segment from Kansas City to New York. Thus, if one of the three links fails, the network management system will automatically select one of the remaining two links without any need for human intervention and without outage noticeable to the user. However, if the failed link is necessary to carry the volume of traffic on the network, users will notice a degradation in response time or service levels.

Additional Comments on Backup and Redundancy

These examples are intended to give you some ideas of techniques you can use in providing backup and redundancy in your data communications network. There are also combinations and variations of these techniques that can be used. The exact approach you select will depend on how important it is for your data communications not to be interrupted, the degree of utilization of your data processing and communications facilities that you desire, and the data communications equipment and software that are available to you either from your computer manufacturer or available to you as an "add on" from other manufacturers.

HIERARCHICAL STRUCTURE AND USER TRANSPARENCY

Intelligent communications network systems, such as those based on SNA or DECNET architectures, have two aspects that are important for you to understand if you are considering using one of these systems—hierarchical layers and transparency.

Hierarchical Layers

The network structure is one of hierarchical layers, each layer consisting of software and hardware with a specific and highly defined role or responsibility. An example of this layered structural approach is shown in Figure 7-9, which depicts five possible layers or levels in an intelligent communications network. Note that the five-level communications control system interfaces with both the user's application program at the top and the physical communications link at the bottom.

You can see from Figure 7-9 that the user's application program accesses the network via the network control level. The network itself is formed around logical channels. You should remember that a logical channel includes the necessary communications control hardware and software to allow data to be transmitted between the two ends of the channel. These logical channels are in turn created out of paths. Paths are the result of taking one or more data links and joining them (if necessary) at intermediate nodes in order to form a route over which the logical channel can be established. Paths are therefore composed of one or more data links. The data link is connected to the data terminal equipment by means of the physical hardware interface such as an RS-232C connector. This

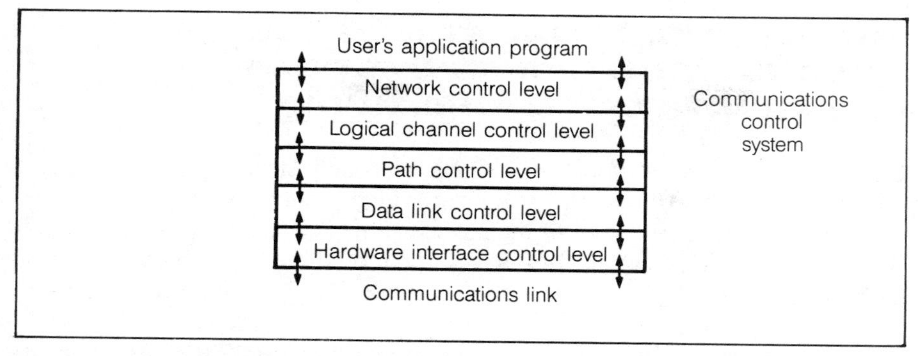

Figure 7-9. A Hierarchical Network Structure

explanation, then, provides an example of the hierarchical nature of an intelligent communications network.

Transparency

The second aspect for you to understand is that each of these layers is a self-contained module—or, in other words, that the internal workings of each of these hierarchical layers or levels are transparent to (that is, they do not affect) the layers above and below and thus to the user as well. Each layer only "sees" the interfaces to its adjacent layers and is not concerned with the internal workings of adjacent or more "distant" layers.

This means that the logical channel level expects the data link level to do the data link level's job, and the logical channel level doesn't worry about what the data link actually is. In the same way, the data link level assumes that the hardware interface level will do its job, and thus the data link level doesn't care what the actual hardware interface is.

For this reason, a network manager can change a segment of a network from a terrestrial (or land-based) circuit to a satellite circuit or can change from a bit-serial hardware interface to a parallel hardware interface without having to make any changes in the logical channel, the network control levels, or the user's applications programs. (These changes, of course, are only possible if the hardware interface level will support both bit-serial and parallel interfaces and if the data link level will support both terrestrial and satellite circuits.)

This independence of the various levels makes the use of an intelligent communications network, which can provide sophisticated redundancy and backup as well as achieve high utilization of all facilities, particularly easy for the average data processing user. It means that you need not be the least concerned about how the network management system is routing the data communications or even about which of several computers on the network you are actually connected to.

THE INTERNATIONAL STANDARDS ORGANIZATION'S OPEN SYSTEMS INTERCONNECTION (OSI) MODEL

Because of the growing importance of hierarchically layered protocols in data communications, the International Standards Organization (ISO) has developed a model called the Open Systems Interconnection (OSI) reference model, which breaks down protocol architecture into seven levels. These seven levels address both network-related protocols and the interface to the applications programs themselves.

The seven levels of the OSI model are shown in Figure 7-10. The OSI model is frequently discussed in data communications articles, and many vendors compare their own layered protocol architectures to the OSI model. You should compare the model shown in Figure 7-9 with the OSI model.

Peer-to-Peer Communications

One of the purposes of layered or modular protocol architecture is to allow two different devices that utilize basically incompatible protocols to communicate with each other as if their protocols were compatible. For example, it allows two computers to communicate with each other even though their physical-layer protocols (that is, those at the hardware interface level) are incompatible. In other words, the network establishes what is called *peer-to-peer communications;* the computers talk to each other in pairs of functions on a peer-to-peer basis.

Here's an example of what this means for the user. In a hierarchically structured network, the DLC functions of each computer communicate with each other without worrying about the types of modems or dialogue management functions being used. The terminal user only worries about the application levels in both computers being compatible without worrying about how to connect an RS-232C to a V.35 interface or about the difference in the computers' data transmission speeds. Each computer is only concerned about the adjacent levels in the model and about the protocol that allows each level to communicate with its peer (that is, with the same function or level in the other computer).

The Seven Levels in the OSI Model

The seven levels in the OSI model (as shown in Figure 7-10) are:

Level 1—Physical Control This level provides the specifications for the hardware (or "physical") interface (such as the RS-232C) between DTE and DCE. Thus, it deals with such items as the network's electrical signals, number of wires, voltages, carrier detection, and so on.

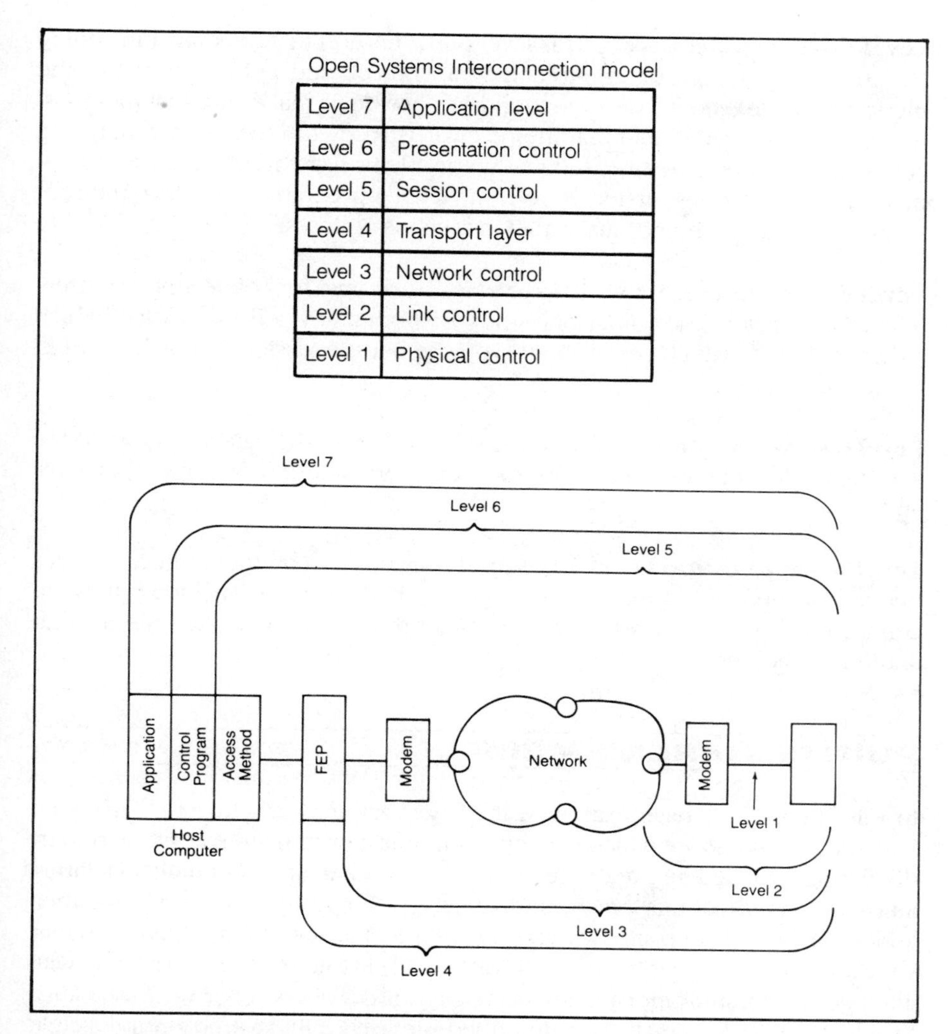

Figure 7-10. The Seven-Level OSI Model

Level 2—Link Control This level handles functions such as frame or block formats, error checking of the frame or block, adding frame-check sequence numbers, and so on. HDLC is an example of a level 2 implementation.

Level 3—Network Control This level handles routing functions, buffering, and so on. The establishment of virtual circuits or the routing of datagrams in packet networks (explained in Chapter 9) are good examples. An example of a level 3 protocol is CCITT's Recommendation X.25.

Level 4—Transport Level This level performs end-to-end control including end-to-end error control, end-to-end flow control, and putting pieces of the message back together. Transportation-layer protocols make sure that the transmitter is ready to send and that the receiver is ready to receive, that there is a connection between the two, and that no problems have occurred during transmisson. After the transmission is over, the level 4 program takes down the connection and lets both ends know the connection is down.

Level 5—Session Control This level is responsible for connecting two users who want to talk to each other once they are both ready; it keeps track of which frames belong to which conversations, of whether the session will be half or full duplex, and so on.

Level 6—Presentation Control This level formats the data for output to the specific terminal being used; it covers codes, format and protocol conversions, and so on.

Level 7—Application Level This level provides the communications services needed by the user's applications programs. These services include setting up connections to the network or to standard program modules like file transfer and remote job entry.

POINTS TO REMEMBER

In this chapter, we have examined the organizational structure of a network. Since you must choose which type of communications line to use, we started by discussing switched and dedicated circuits for both analog and digital facilities. Since dedicated circuits can be out of service for hours if they fail, we then looked at how dial-up circuits can be used as backup for dedicated circuits. Various forms of redundancy were discussed, including those available with intelligent communications networks such as SNA, DECNET, or various packet-switched networks. Finally, we described hierarchically layered protocols and the seven-layer OSI model.

The basics of network architecture and access methods covered in the previous chapter and the concepts of network organization we have covered in this chapter will enable you to discuss with vendors the specific network approaches available for the data processing or data communications equipment and software you are considering. With this background information, you can make meaningful comparisons of the data communications network options available to you.

8

REDUCING DATA COMMUNICATIONS COSTS AND TRANSMISSION ERRORS

As your use of data communications increases, you will certainly want to find ways of reducing your data communications costs and your transmission errors. Two of the major cost elements of data communications are the cost of the transmission lines and the cost of the access ports to the computer. In this chapter, we look at how you can reduce line and port costs and then at how you can reduce errors that are generated during transmission of the data.

REAL-LIFE EXAMPLES OF THE NEED TO REDUCE COSTS AND ERRORS

The following are examples of common data communications situations that raise questions about cost and error reduction. The techniques discussed later in this chapter will provide answers to these questions.

1. As the use of personal computers and computer terminals grows, you will probably find a demand throughout your company to connect terminals in offices to computers at remote locations (such as at your data center or in other offices). Running private data lines from all of the remote terminals to your computer center can be very expensive. How can you accommodate user demands at the lowest possible costs?

2. Your company publishes catalogues, and you would like to transmit the text of the catalogue directly to the typesetter, who will give you a lower price for doing so. However, you are concerned that transmission errors will go undetected since your computer has only simple parity error checking. Of course, you cannot afford to have errors in the catalogue caused by the data communications process. What can you do?

3. Your company is considering using distributed data processing with the CPUs, the data base, and the user terminals situated at many different locations.

Running separate data lines to all the possible combinations of connections just isn't practical. What is your alternative?

4. A new office is opened thirty miles away from the main computer. The office manager needs eight computer terminals for a customer-service computer inquiry function. When you tell him the cost of running eight data lines to these terminals, he asks, "Isn't there any way to do it at a lower cost?"

5. You've received a request to add eight computer terminals to your computer system; however, you only have two ports left on the computer, and it will be six months before you can get more ports. When you explain this to the department manager, she says that the terminals will be used very infrequently but they are critical to her new department—isn't there something you can do more quickly?

6. Your company is headquartered in San Francisco, which is where the corporate computer is. Management decides to put computer terminals in the Fresno, Los Angeles, Riverside, Orange County, and San Diego offices, but the top executives go through the roof when told what separate data lines to each location will cost. Since your computer is a mini without sophisticated data communications capability, you cannot connect directly more than one terminal to each computer port or data line. What can you do to reduce costs?

THREE TECHNIQUES FOR REDUCING COSTS AND ERRORS

The above situations can all be addressed by one (or perhaps by a combination) of three approaches:

1. Sharing communications lines, so that two, three, or more terminals or devices are all connected to a single line—this eliminates the need for and cost of a single line for every device.
2. Sharing computer access ports, so that two, three, or more terminals or devices are all connected to a single computer port—this eliminates the need for and cost of a single port for every device.
3. Using sophisticated error detection and correction schemes to reduce errors introduced during transmission to a minimal level, even if your computer is not designed to utilize those techniques.

How can these solutions be implemented on a computer system that was not designed to utilize these cost- or error-reduction techniques? In the balance of this chapter, we will discuss how separate "add on" devices or "black boxes" can make these three cost- and error-reduction approaches possible.

HOW TERMINALS CAN SHARE COMMUNICATIONS LINES

In Chapter 6 we discussed how a point-to-point or star network with each line connecting only one computer terminal to the host computer can be very expensive in terms of the costs of both communications lines and computer ports (see Figure 8-1). Sophisticated mainframe computers overcome this cost problem by using polled communications techniques where multiple terminals are connected to a single communications line and each terminal's messages are preceded by that terminal's unique address (we also talked about polling in Chapter 6).

However, polling techniques, in general, are not directly available to users of most mini- and microcomputer systems, since those systems historically have been designed to utilize "dumb" computer terminals ("dumb" from a data communications standpoint). Let's look at the differences between dumb and intelligent terminals.

Dumb Versus Intelligent Terminals

The "dumb" and "intelligent" terminal terminology is another one of the confusing areas in data communications. There are many different meanings associated with the term "dumb terminal." For example, if a terminal has the capability to edit what is typed into it or to format the CRT screen in different ways, it may be referred to as an intelligent terminal, yet, for data communications purposes, it may still be a dumb terminal.

For the purposes of data communications, a dumb terminal cannot:

- Recognize an address, so it cannot be polled or selected
- Perform any kind of error control or detection
- Operate synchronously
- Work with a communications protocol other than the simplest character-by-character async protocol
- Transmit in a formatted message mode
- Reformat data it receives (it must display or print data exactly as it is received)
- Add block check characters, control characters, or sequence numbers to its output
- Operate in half duplex with line turnaround (dumb terminals with an HDX/FDX switch usually use the HDX setting for a local copy mode—that is, whatever is typed shows up on the screen or paper—and the FDX setting for echo-check mode), although they may be able to operate half duplex on a four-wire line since no line turnaround is required.

From a data communications standpoint, intelligent terminals can do at least some of the above items. In the literature, you may see terms like "semi-intel-

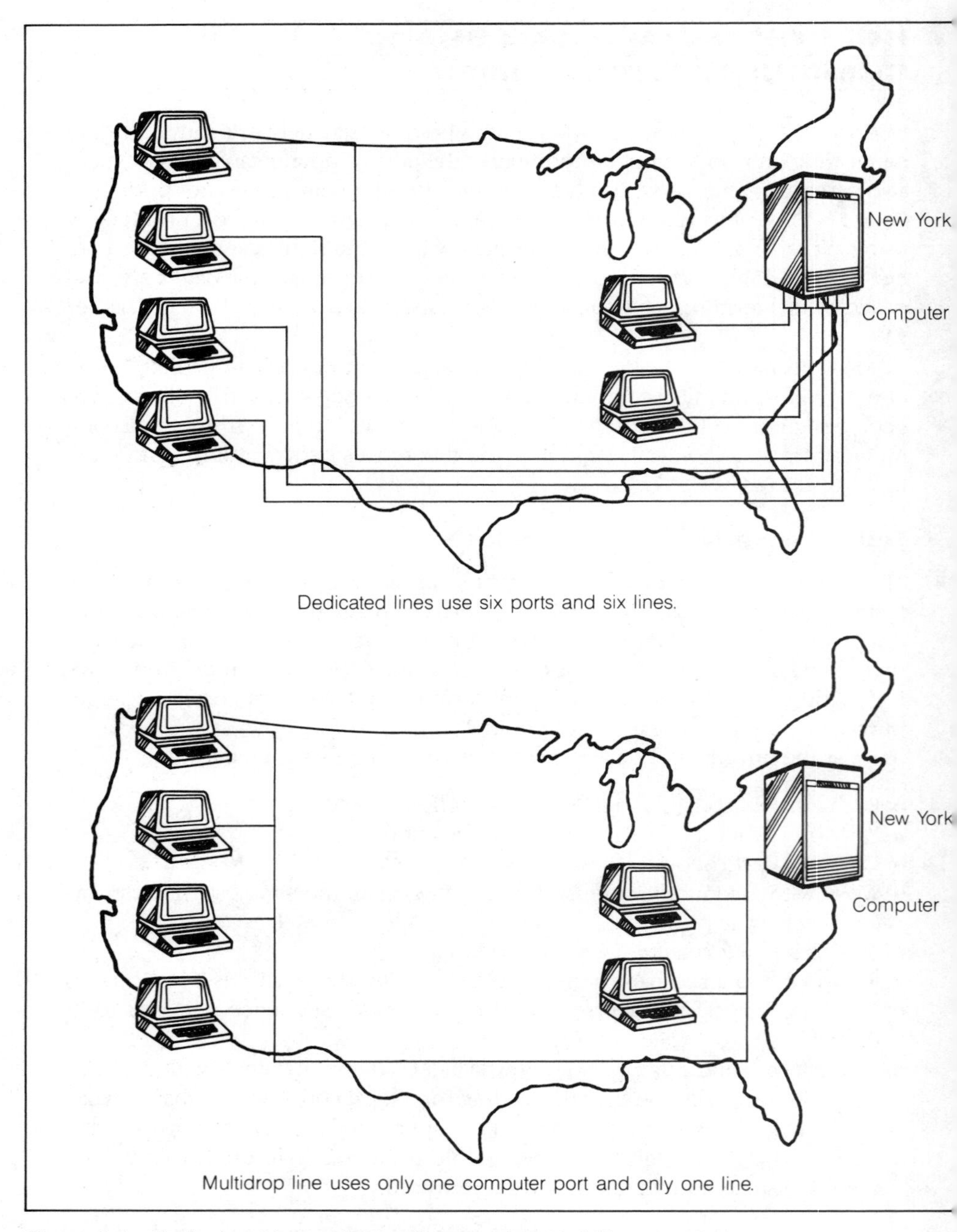

Figure 8-1. Reducing Line and Port Costs with a Multidrop Line

ligent" or "smart dumb terminals." Personal computers, home computers, and workstations are also terms you will hear applied to devices that can be intelligent terminals if they are equipped with data communications hardware and software.

Don't be overly concerned about the meaning or usage of these terms. They will vary depending on who is using them. Remember the limits of dumb terminals and be sure the terminals that you are considering will perform the data communications functions you need.

In the sections that follow, we talk about how you can add "black boxes" to your system that can add "smarts" to a dumb terminal.

Multiplexors

Multiplexors divide a standard voice-grade circuit into two or more data communications channels. In this way, two or more "dumb" terminals can share the same physical circuit (see Figure 8-2). Multiplexors are simply black boxes that combine or mix a number of channels onto one circuit by electronic means.

There are two basic types of multiplexors: frequency division multiplexors (FDM) and time division multiplexors (TDM).

Frequency Division Multiplexors Frequency division multiplexors are analog devices that simply divide the transmission bandwidth (that is, the frequency range) available on the circuit into smaller bandwidths and then allocate one of those smaller bandwidths to each data communications channel. For example, a standard voice-grade line can easily accommodate a signal 2400 Hz wide. However, a dumb computer terminal transmitting with a signal 300 Hz wide would tie up the complete circuit if connected with a conventional modem (see Figure 8-3). An FDM can divide the 2400 Hz bandwidth into six 300 Hz channels (plus guard bands separating each channel), and thus the same circuit that could support only one terminal can now support six terminals, significantly lowering communications line costs.

Frequency division multiplexing is an analog technique. Therefore, computer terminals are connected to FDMs through modems (which convert the computer's digital signal to analog), and the FDMs themselves are connected directly to the analog telephone line.

The FDM technique is somewhat inefficient since, among other things, it does not utilize the entire bandwidth for transmission (unused guard bands are required to separate channels). However, historically FDM has been less expensive than time division multiplexing (TDM). Now, with improvements in digital technology, FDM's cost advantage has disappeared, and TDM is the predominant technique used with circuits like those provided by the telephone company. Nevertheless, FDM is widely used in data communications systems based on

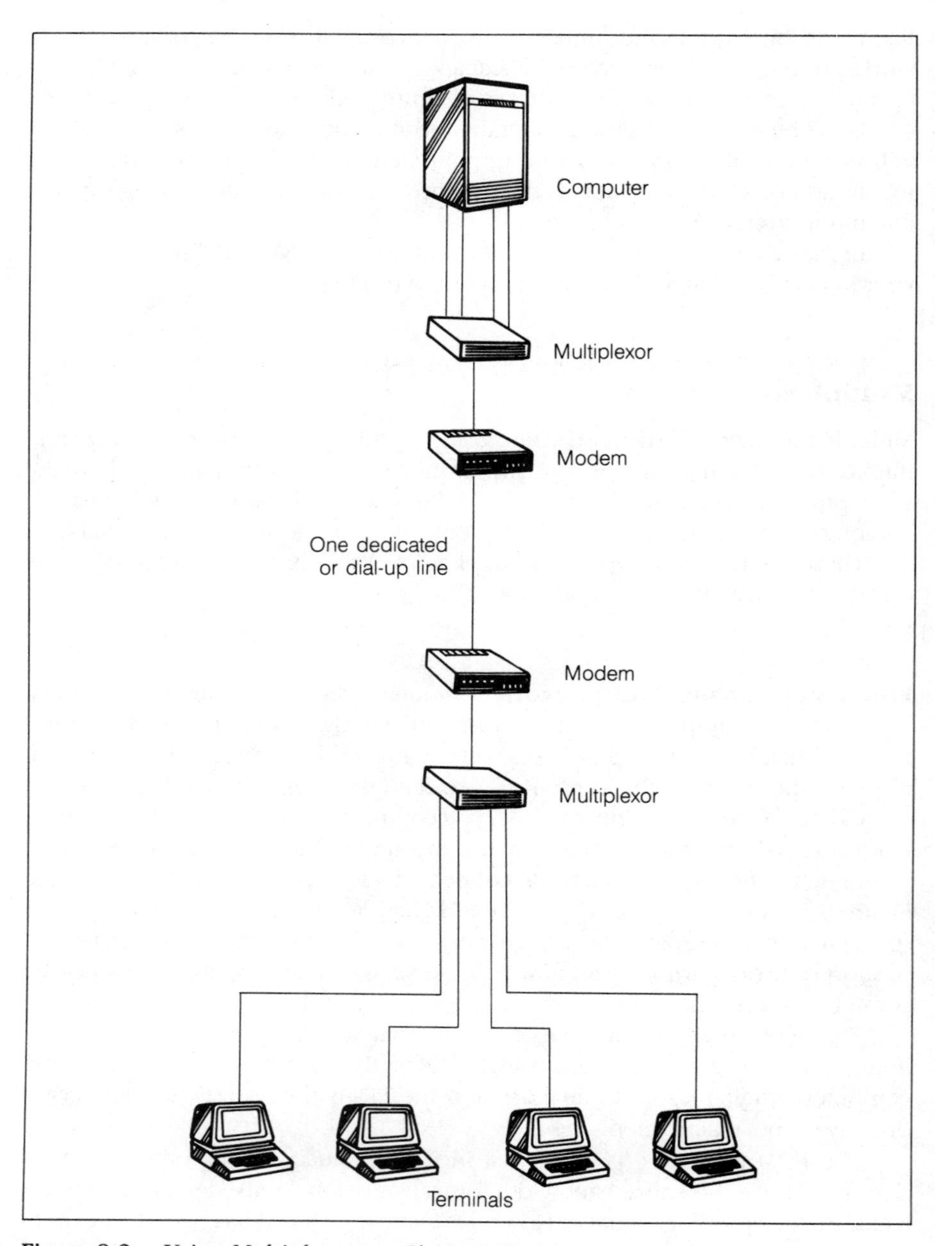

Figure 8-2. Using Multiplexors to Share a Circuit

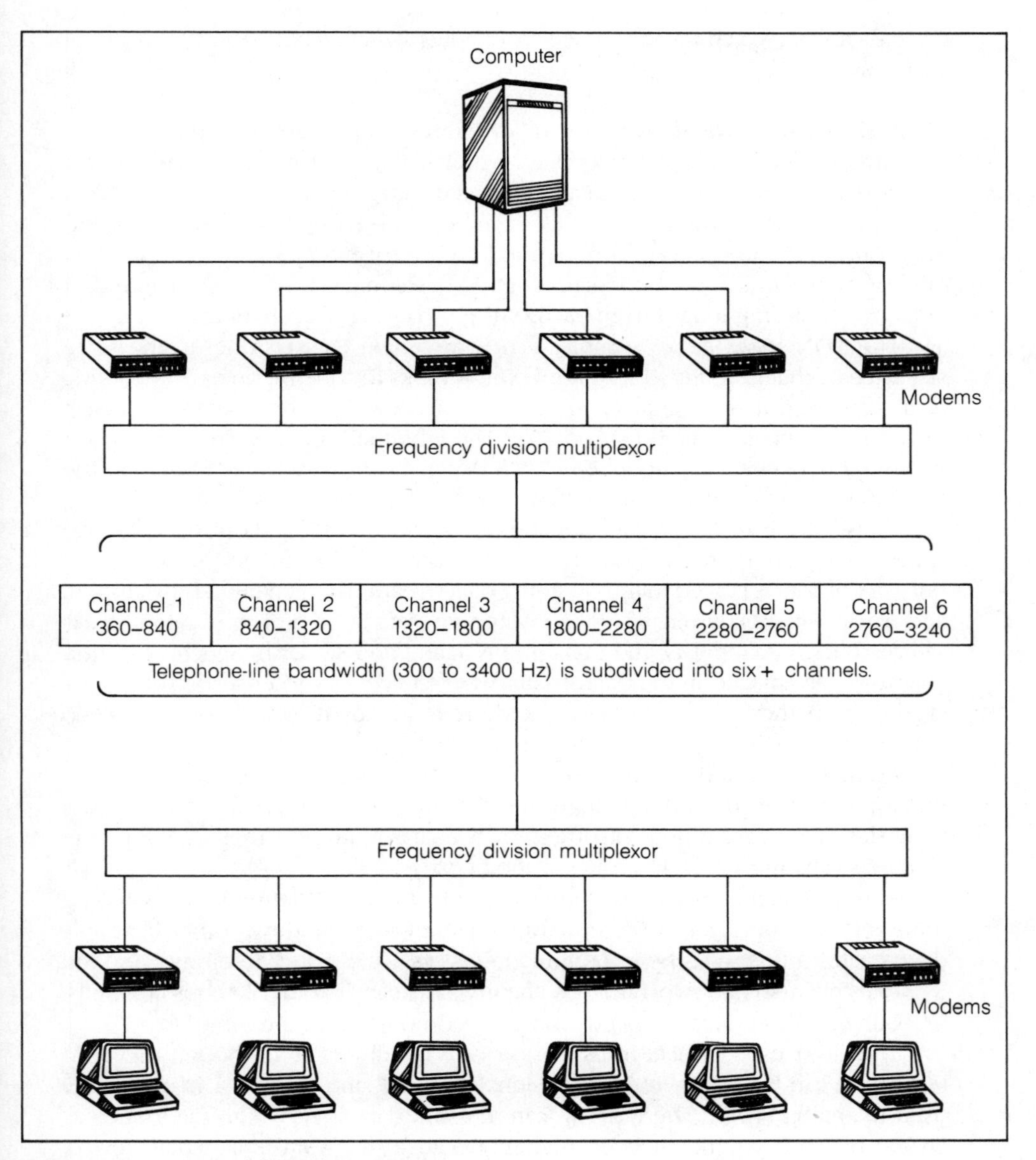

Figure 8-3. Frequency Division Multiplexors

coaxial cables (which are discussed in Chapter 9); Ethernet is one example of such a system.

Time Division Multiplexors Time division multiplexors are digital devices. The time division multiplexor creates a digital data stream that it will use for transmission purposes and breaks the stream into units (or samples or slices) usually of uniform length. These slices can be one bit long (called *bit multiplexing*), one byte or character long (called *character multiplexing*), or of some other arbitrary but usually uniform length. These time slices are filled with data of a corresponding length from each of the incoming data channels in turn.

The TDM looks at or "samples" each incoming data channel in the same size slices—that is, if it is a bit multiplexor, it looks for one bit on each incoming data channel; if it is a character multiplexor, it looks for one character on each incoming data channel; and so on. The TDM then takes a data slice from each incoming data channel and assembles those slices in a specified order on the outgoing transmission data stream (see Figure 8-4).

Thus, if four incoming data channels are to be multiplexed onto one higher-speed data channel, then Figure 8-4 shows how first channel A is sampled and one unit of data is placed in the outgoing data stream. Then channel B is sampled and a unit of data is placed in the outgoing data stream. Then channel C is sampled, then channel D, then channel A again, and so forth. Synchronization characters are placed in the data stream where necessary to ensure proper synchronization; they tell the multiplexor where to line up the bits in order to keep track of which data go with which channel.

At the far end of the multiplexed circuit, the process is reversed. The transmission data stream is broken apart into time slices, and the bits of data in each time slot are reassembled into their own data streams on the outgoing data channels (channels A, B, C, and D in the figure).

Although Figure 8-4 shows one sample from each channel in an outgoing frame of data, the TDM can sample one channel twice, and two other channels once each. In this way, the same four time slots are used to accommodate one 2400 bps channel and two 1200 bps channels instead of four 1200 bps channels. Of course, other mixes of speeds can be accomplished in the same way.

The TDM used in data transmission is a digital device (although the TDM technique can be used with analog signals as well, and, in fact, it is frequently used in analog PBXs). Therefore, it can transmit data at very high rates, such as 56,000 bits per second or even higher. Because the TDM is a digital device, computer terminals are connected directly to the TDM and the TDM is connected to the analog telephone line through modems. Contrast this to the FDM connection described earlier and look again at Figure 8-4.

The TDM is efficient and flexible, but it is limited by one of its basic concepts: once a pattern has been established as to which time slots on the main transmission channel belong to which incoming data channels, those time slots are dedicated to that channel whether the channel is transmitting data or not. Since

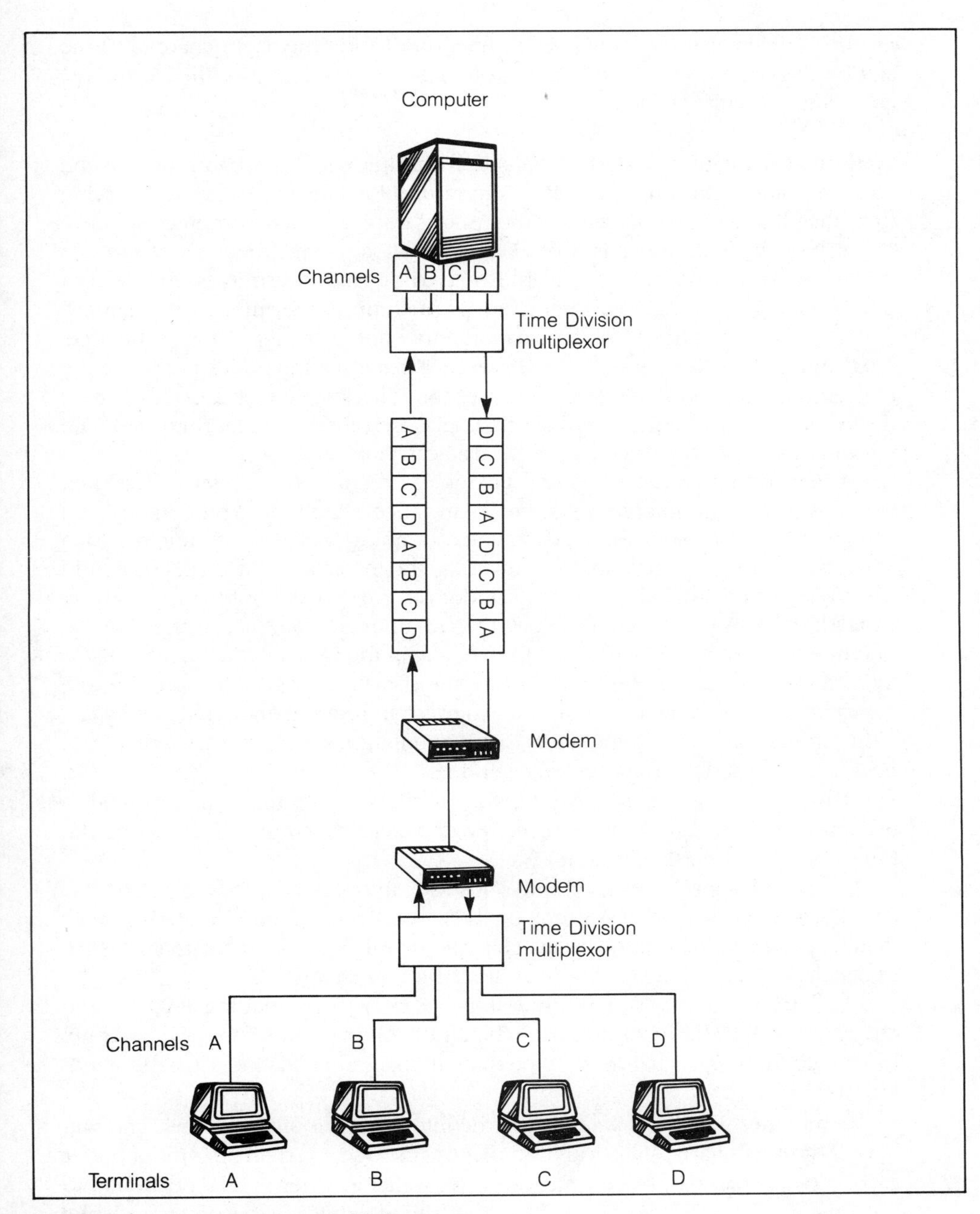

Figure 8-4. Time Division Multiplexors

most terminals transmit only a small fraction of the time, each channel's time slot on the main transmission stream often goes unused and is filled with PAD characters or with idle characters.

Statistical Multiplexors The statistical multiplexor is probably one of the most important innovations in data communications for the user who is trying to reduce line costs, especially for the user of mini- or microcomputer systems. The statistical multiplexor is a black box that is also known as a *stat mux* or intelligent TDM. The stat mux enables you to put more terminals on one data line by taking advantage of a characteristic of computer terminal use—namely, that terminals typically utilize their data channel only a small portion of the time. A stat mux works like a standard TDM except that a stat mux does not waste the unused transmission time; instead, it uses the idle time slots that occur when a device is not transmitting data. By utilizing these otherwise idle time slots, the stat mux increases the circuit's overall data-carrying capacity.

A very rough analogy that will give you a feeling for the way a stat mux works is that of the intercom system on an office telephone. You could set up twenty buttons on your phone, each corresponding to a separate line for each of the twenty people you wanted to talk to on the intercom. However, you would use any one of those twenty lines so seldom that the cost wouldn't be worth it. Instead, you can install one line that goes to all twenty people, connecting the one line to a button or dial. You can then signal the person you want to talk to by means of a buzzer or bell. In this way, the conversations to all twenty people are carried on the same one line, since they are infrequent enough that they can all fit on that one line. Remember, that if twenty lines were used, there would be a lot of unused or idle time on each line.

The stat mux works in a similar way. It allows more than one terminal to use a time slot by signaling the receiving end as to which terminal is using the time slot for that particular cycle.

One of the approaches used with statistical multiplexors is to label or address each slice of data sent. The stat mux on the far end, then, knows by labels (rather than by position) the outgoing data channel to which it should assign each slice of data as the data come off the main transmission channel.

This means that very few time slots are wasted in sending PAD or idle characters when the incoming data channel is empty. Therefore, the stat mux can create more data channels on the main transmission channel than the standard TDM.

Sometimes when you want to use the intercom, it is already in use and you have to wait. Similarly, statistical multiplexors include a certain amount of buffer capability so that, if no idle time slots are available when a channel wants to transmit, the data are stored or buffered until an idle time slot becomes available.

Since a standard TDM assigns one time slot to a channel, a TDM can compress four 1200 bps channels onto one 4800 bps main transmission channel (4 ×

1200 = 4800). However, by combining the use of idle time slots and buffering, a stat mux can compress six, eight, or even twelve 1200 bps channels onto one 4800 bps main transmission channel.

Remember, the reason a stat mux can get eight to twelve 1200 bps channels (12 × 1200 = 14,400 bps) on a 4800 bps circuit is that data are not being sent continuously by all twelve devices at 1200 bps. Data are being sent in short bursts at 1200 bps when they *are* being sent, but data are in fact being sent only a small percentage of the time. Therefore, a two-to-one or even better data compression ratio is possible with a stat mux, depending on what percentage of the time the incoming data channels are actually used.

Variable-Length Time Slots Since the stat mux's primary role is to pack several data channels on one circuit, the manufacturers continue to look for ways to increase data throughput with stat muxes. The technique described above, where each slice of data is labeled, creates nonproductive overhead. That is because the bits used as labels take up data transmission capacity. And the process of labeling itself reduces data throughput. If there was a way to utilize idle time slots and not have to add labels, data throughput could be increased even more.

Therefore, another approach was developed and is used in multiplexors such as the Codex Corporation's series 6000 intelligent network processors. This approach does not label each time slot but rather uses variable-length time slots. An idle channel is assigned a time slot of minimum length. An active channel is assigned a time slot with a length equal to the number of characters it has to send, up to the maximum number allowed per slot. This technique, coupled with data compression, can consistently produce overall data compression ratios of up to four to one.

Data Compression

Start/Stop Bit Stripping Standard "dumb" communications devices use asynchronous transmission with start and stop bits on each character. If one start bit and two stop bits are used for each eight-bit data character, then three out of every eleven bits transmitted, or 27 percent, are overhead and not data.

The stat mux and often the standard TDM can strip the start and stop bits off asynchronous transmissions and replace them with less frequent sync characters. Thus, the stat mux needs to transmit only eight bits plus an occasional sync character, where ten or eleven bits were needed for each async character if transmitted without the bit stripping. This compression helps compensate for the overhead involved in the time-slot labeling schemes used in conventional stat muxes.

Repetitive Character Coding Another example of data compression is to look for a long string of the same character. For example, dumb terminals often send a whole line of blanks. One approach to data compression is to recognize this string of blanks and send a message such as "*40 ". This message is interpreted by the stat mux at the receiving end as being equivalent to a string of forty blanks, but only four characters have to be sent instead of forty.

The use of data compression techniques such as bit stripping and repetitive character coding not only compensate for the overhead introduced by the multiplexing technique but also allow an even greater data throughput on a single circuit. In this way, a single line can accommodate either more devices or higher-speed devices than would otherwise be possible.

Thus, various types of multiplexors and data compression devices are available in the form of black boxes that can be added to both ends of a data line and provide an alternative to the expense of maintaining one communications line per terminal or other device. Figure 8-5 provides a schematic comparison of the different combinations of multiplexors, modems, and data compression units.

Variations on the Multiplexor

Inverse Multiplexors Until now we have looked at using a multiplexor to combine two or more data streams (such as the data streams from two or more terminals) onto one data circuit. However, you might have a data stream that is too broad to fit on a voice-grade circuit, say a stream of 19.2K bps. Perhaps you do not want to pay for a wide-band facility when you only need this high-speed circuit one or two hours a day and you already have two voice-grade circuits with 9600 bps modems on them with several spare hours a day available. In this case, an *inverse multiplexor* can be used to divide the 19.2K bps data stream into two 9600 bps data streams, each of which can then be transmitted over one of the 9600 bps data channels you already have available (see Figure 8-6).

Concentrators The statistical or intelligent multiplexor is referred to as a concentrator by some vendors. However, the device more traditionally referred to as a concentrator has different characteristics than a stat mux, although there are many similarities between the two.

The traditional concentrator usually is installed only at one end of a circuit. The other end is connected directly to the computer (or to the front-end processor) through a connection that is designed to accept data directly from the concentrator. The concentrator usually provides store-and-forward capability on a full message basis while most intelligent muxs work on a character or bit basis. Also, the concentrator might perform additional functions that are not normally handled by a stat mux, such as selective routing, editing, buffering long messages, data compression, and forward error correction.

The IBM 3276 cluster controller, used with such 3270 CRT terminals as the 3278s, is an example of a commonly used concentrator. Several 3278 CRT ter-

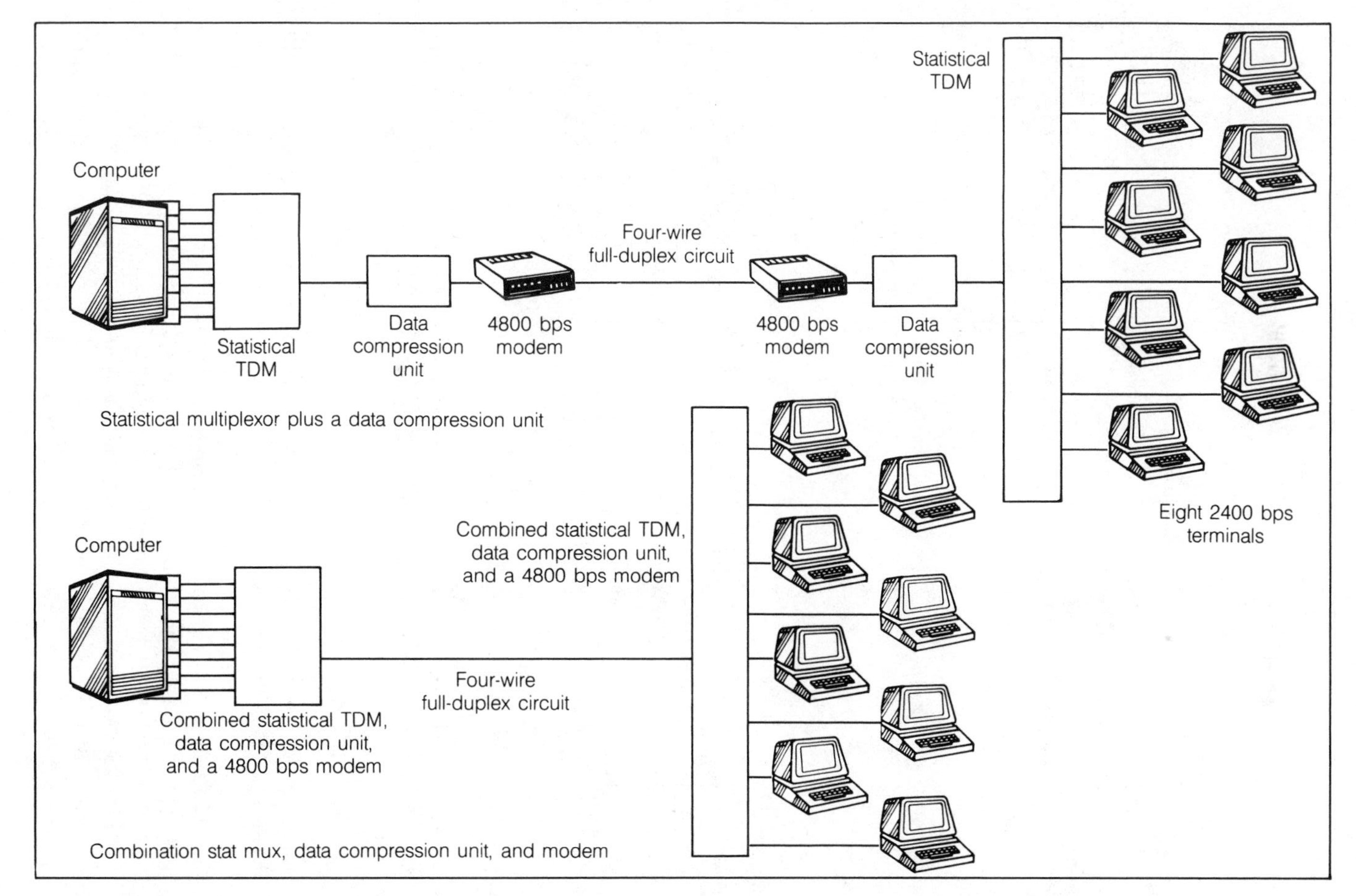

Figure 8-5a. Examples of Multiplexors, Modems, and Data Compression Units in Combination

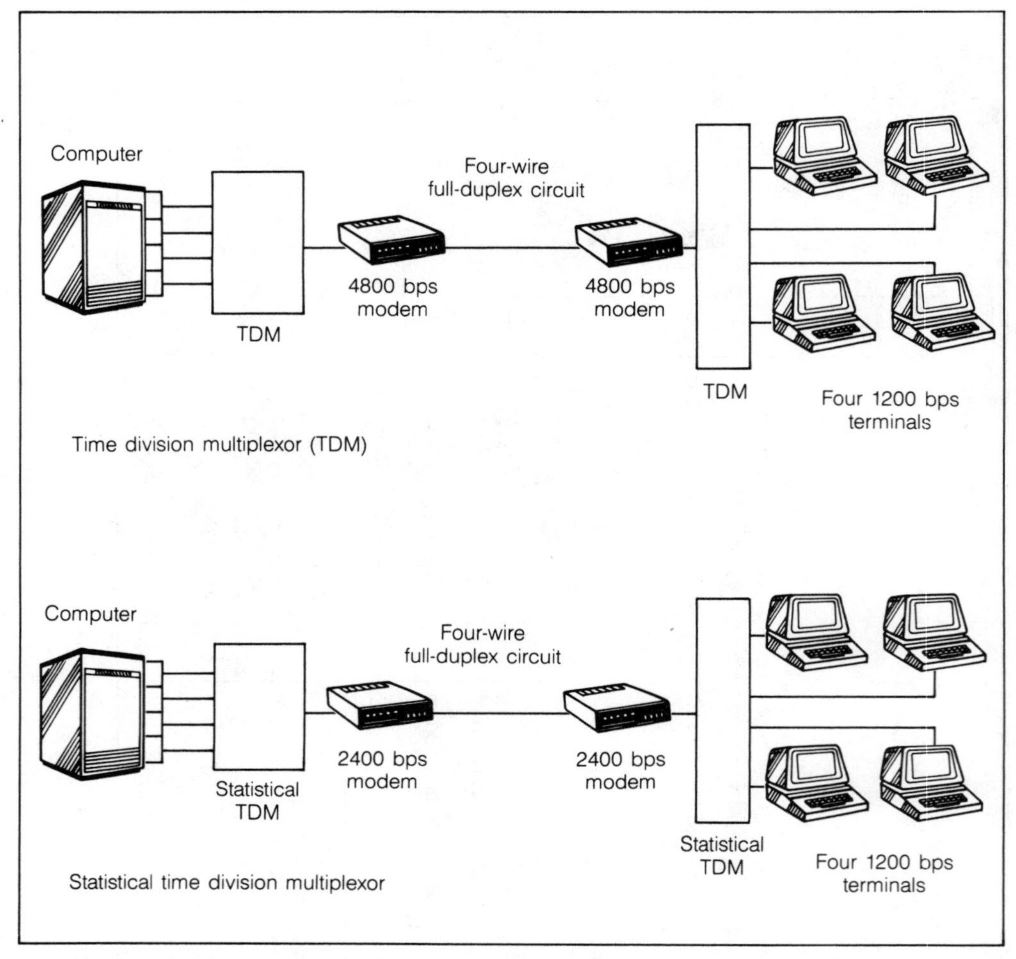

Figure 8-5b. Examples of Multiplexors, Modems, and Data Compression Units in Combination

minals are connected to the 3276, which then connects to a single port on the IBM host computer (see Figure 8-7).

The Multidrop Multiplexor Some intelligent multiplexors can function in a multidrop mode. In this environment, one mux serves as the master and the other muxes serve as remotes. In this arrangement, the intelligent multiplexor provides the polling function for the multidrop line and thus allows dumb terminals to be multidropped either singly or in clusters in conjunction with a computer that was not designed to support multidropped terminals.

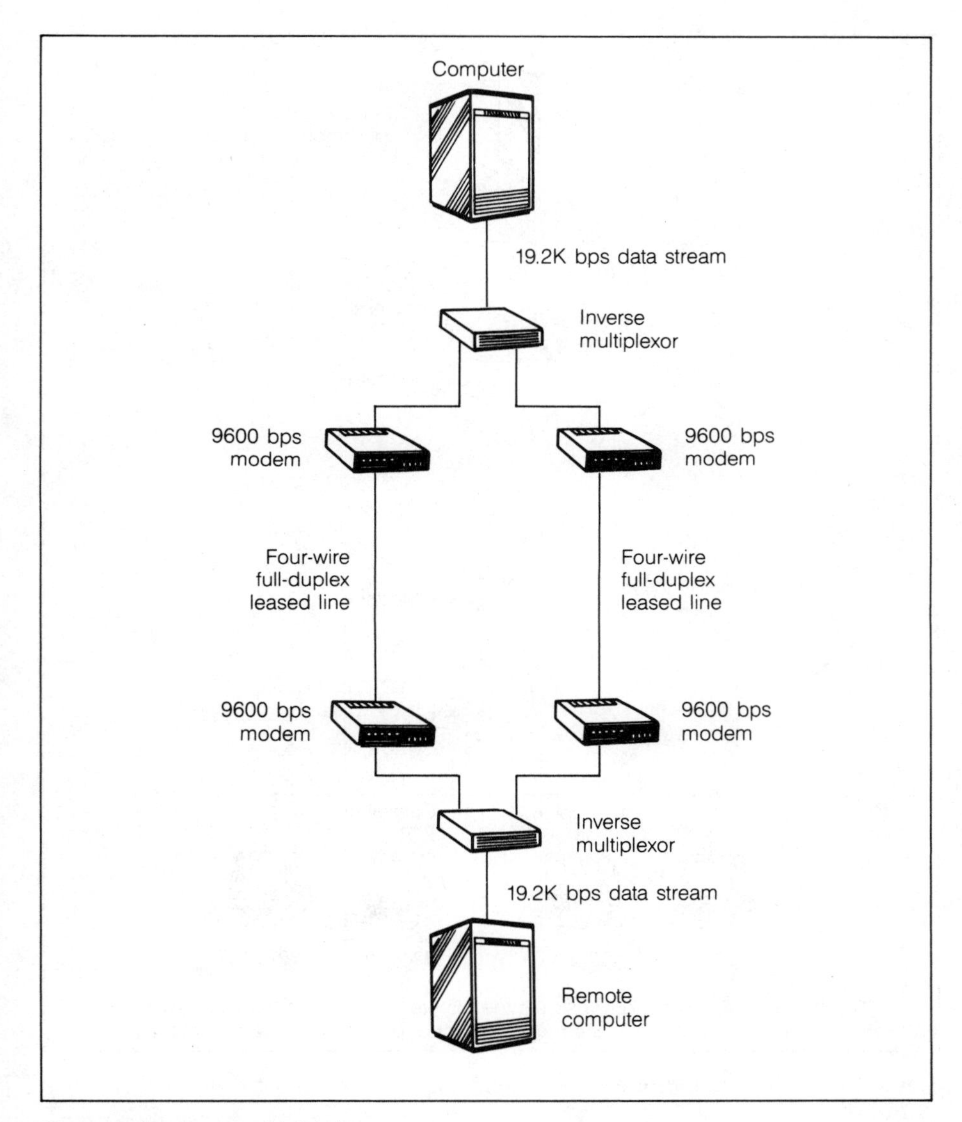

Figure 8-6. Inverse Multiplexors

Example 6 at the beginning of the chapter, with the computer in San Francisco and terminals located in five other offices in central and southern California, describes a situation where this approach can be used (see Figure 8-8).

Frequency division multiplexing can also be used to multidrop terminals on one circuit, but the digital stat mux allows the use of much higher speeds than does the FDM on voice-grade lines. Thus, FDM is usually only considered

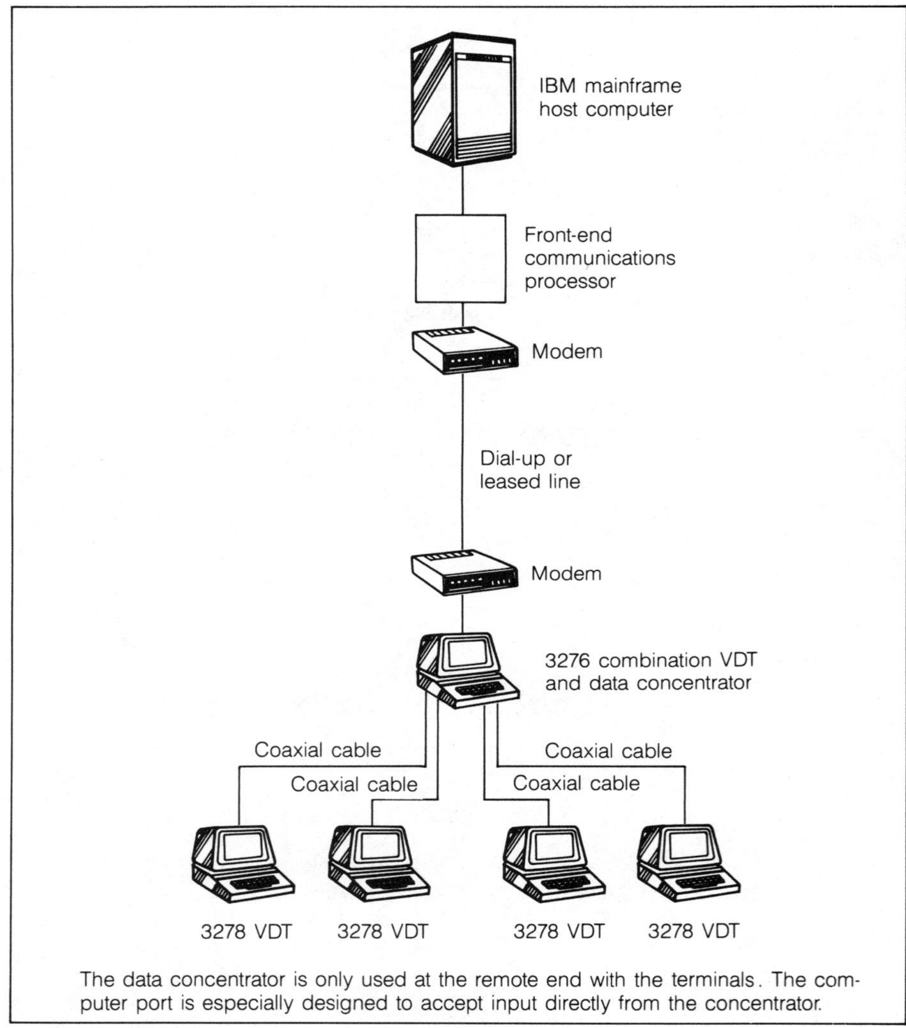

The data concentrator is only used at the remote end with the terminals. The computer port is especially designed to accept input directly from the concentrator.

Figure 8-7. An Example of a Data Concentrator

for slow-speed applications such as 110 bps or 300 bps or for coaxial cable applications.

REDUCING COMPUTER PORT COSTS

In addition to line costs, the other major data communications cost is the number of computer ports used. Unless a computer can support multiple terminals on one port, each time a terminal is added, a computer port must be added. On

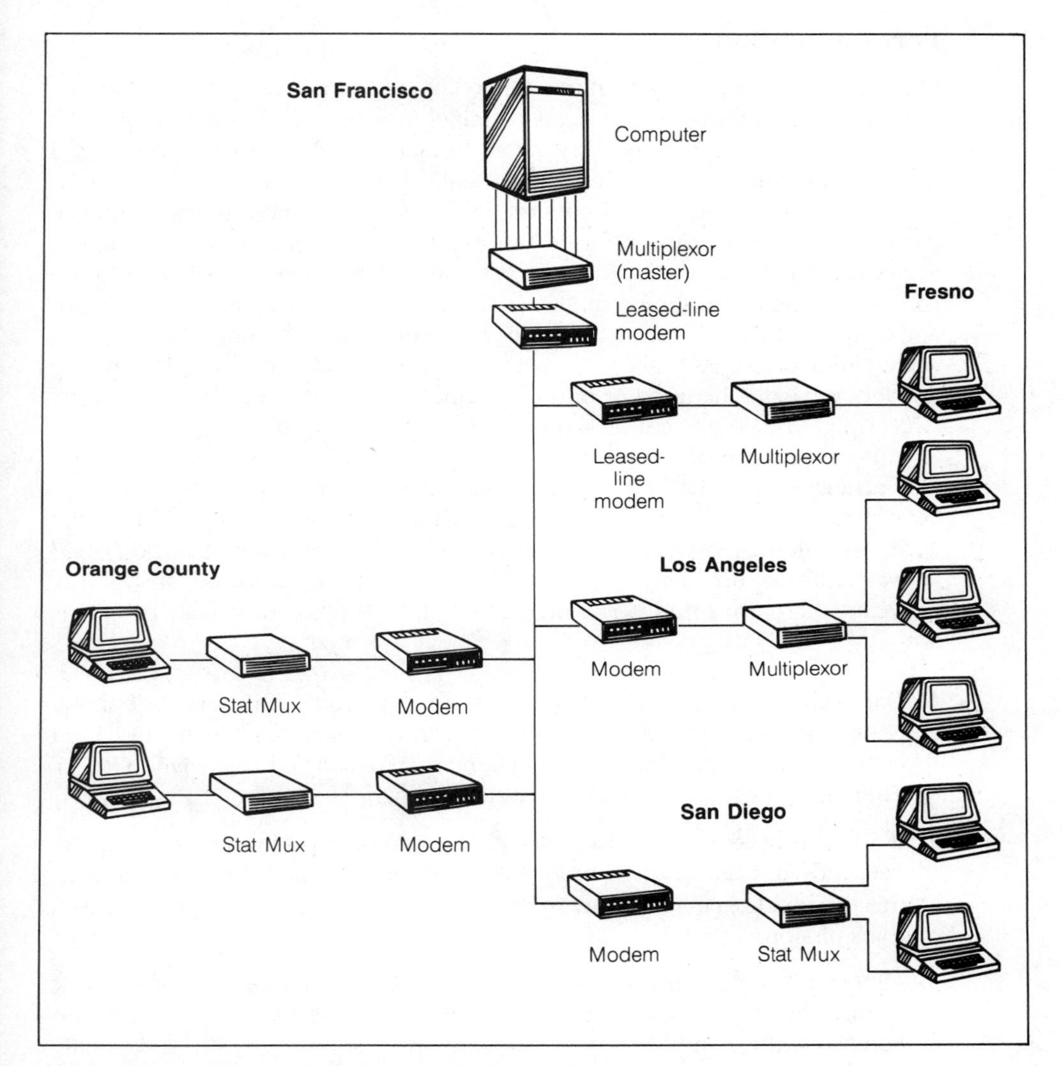

Figure 8-8. Using Multiplexors on a Multidrop Line

many computers, ports are very expensive, and in most cases computers can only have a limited number of ports.

This means that the $500 CRT terminal you want to add may cost $5,500, since another $5,000 access port is needed on your computer. Even worse, you may find all of the ports on your computer are in use, and no more ports can be added. In that case, you can't add your terminal unless you are willing to disconnect another terminal from your computer.

The following subsections describe a number of different approaches to solving these problems and reducing port costs.

Port Contention

One approach to reducing the number of computer ports required is *contention*. Contention is a system where many terminals are not permanently connected each to their own computer port; rather, they contend for access to a limited number of computer ports when they actually want to use the port.

This contention approach is frequently used with remote timesharing systems. With these systems, the user's terminal is not permanently connected to the computer. When the user wants to connect a terminal to the timesharing computer, the user dials the computer's telephone number, the computer answers, and connects the user's terminal to one of the available ports. Since the terminals do not all need access to the computer at the same time, only enough ports are needed to handle the number of users connected at the period of peak activity, which might be 50 percent or less of the total number of users.

The most commonly used port contention method is to use telephone numbers arranged in a rotary sequence or "hunt group" (either telephone company central office numbers or local PBX extensions—see Figure 8-9). This is the approach often used with remote timesharing systems such as that described above. While dial-up hunt-group contention systems have been used satisfactorily for years, there are a number of problems with them. Some of these problems are:

1. Data calls tend to be longer in duration than voice calls, and telephone companies usually charge for most calls from businesses (even most local calls) by the length of the call (most contention systems do use dial-up lines). Therefore, these dial-up calls can be expensive.

2. If a PBX is used to switch the calls, PBX switching capacity is used up by the data calls and can result in more PBX equipment being required. However, this may not be a major problem with modern digital PBXs designed for data transmission.

3. All terminals, even local ones, must be connected by modems if switched by a voice PBX or a voice dial-up line. Note that data PBXs or digital lines can be used without modems; however, some type of line access unit is usually required.

4. Since modems are used on dial-up central office lines or with a voice PBX, data transmission speeds will be limited both by the speeds achievable with conventional modems and by the cost of high-speed modems. Another limiting factor may be the speed of data transmission that can be handled by the PBX itself without introducing distortion (usually 2400 to 4800 bps with electronic PBXs designed for voice traffic). Again, this is not usually a problem on data PBXs or digital PBXs specifically designed with data transmission in mind.

5. The PBX or telephone company central office may not have the ability to keep trying automatically to establish the connection if all ports are busy.

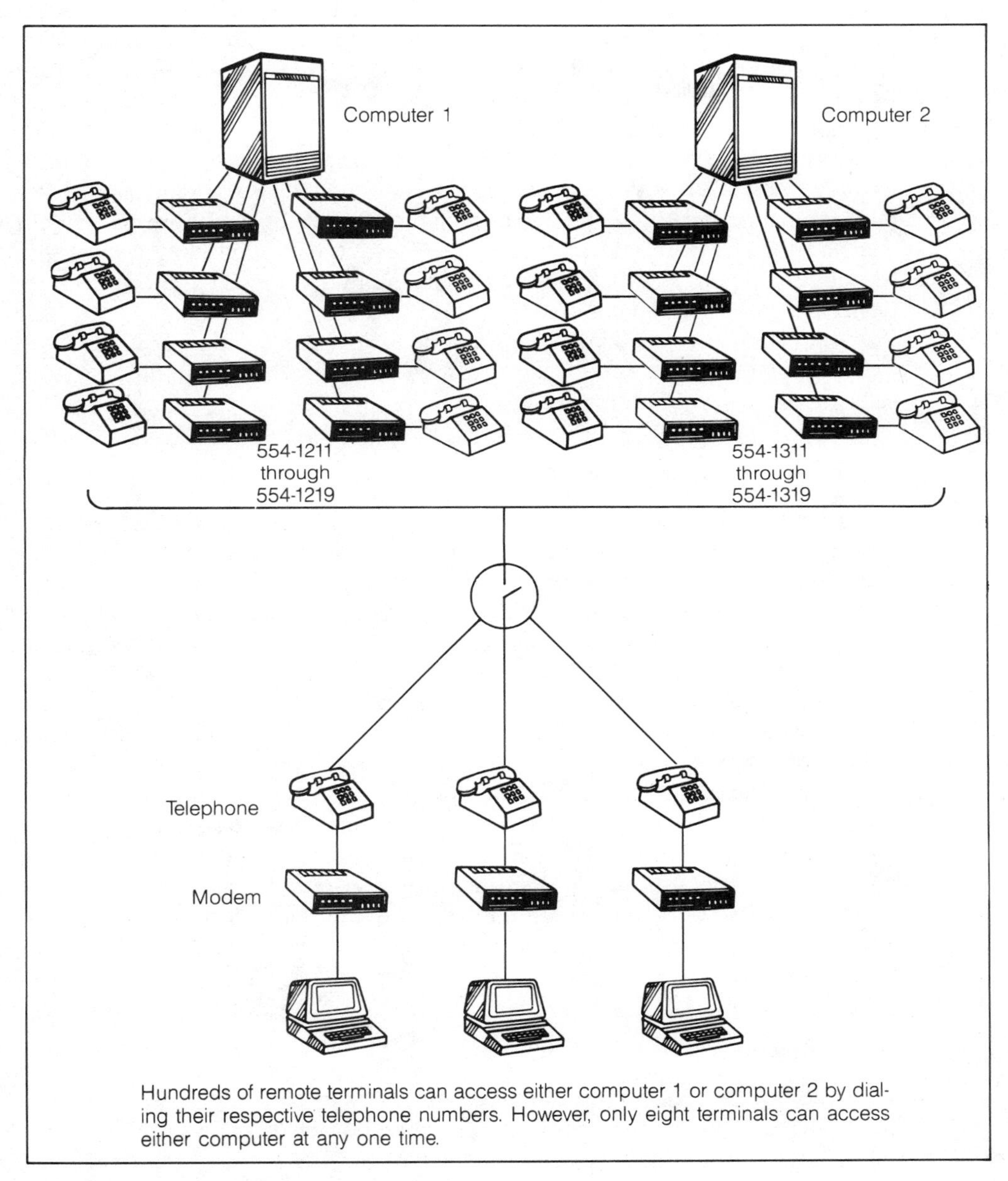

Figure 8-9. Port Contention

Port Selectors

Microprocessor-based black boxes known as *port selectors* overcome many of the problems associated with port contention. Among other features, port selectors allow a user to connect to both dial-up and dedicated lines. Furthermore, these lines can be either analog lines connected by a modem or digital lines connected directly to the port selector. These lines can be mixed in a variety of combinations while allowing the port selector to perform the same function as port contention (see Figure 8-10).

Port selectors are operated from the terminal keyboard or by the computer wishing to establish the connection. By transmitting the appropriate characters, the terminal or the computer signals which computer or device it wants to be attached to. The port selector can determine automatically the speed of the transmitting device and thus the correct group of ports to which to connect the device on the requested computer.

For a company with a voice PBX that does not have queuing or a company that has a need to connect dedicated short-haul or current-loop circuits, the port selector offers significant advantages.

A Port Selector as a Data PBX

The growth of electronic mail and messaging has made users more aware of the benefits of being able to communicate not just between a terminal and a host computer but between two terminals as well. Also, distributed data processing installations, such as that described in example 3 at the beginning of this chapter, make it desirable to be able to select among several different computers, data bases, or terminals that a user may want to communicate with. A data PBX—that is, a PBX designed to switch data rather than voice telephone calls—is one answer to these problems, but traditional data PBXs are very expensive for small installations. Newer digital PBXs designed for voice and data may be cost effective in these situations.

Also, the port selector can also function as a data PBX. Rather than connect a terminal to one of several ports available on a computer, the port selector can be used to connect the terminal to any other specific terminal or computer connected to the port selector. In this situation, the port selector works in the same manner as a voice PBX when it connects one telephone to another telephone attached to the PBX.

Port Concentrator

If, in addition to reducing the number of data lines, you want to reduce the number of computer ports required but want all of the terminals connected to the computer at all times, a black box called a *port concentrator* can be used.

Remember that the statistical multiplexor concentrates the data being sent to and from several devices so the data can all fit on one transmission line. The far-end multiplexor then breaks the transmission data stream down into multiple channels so that each channel can be connected to its own computer port.

The port concentrator can be substituted for the stat mux at the computer end of the circuit. There, the port concentrator performs many of the functions

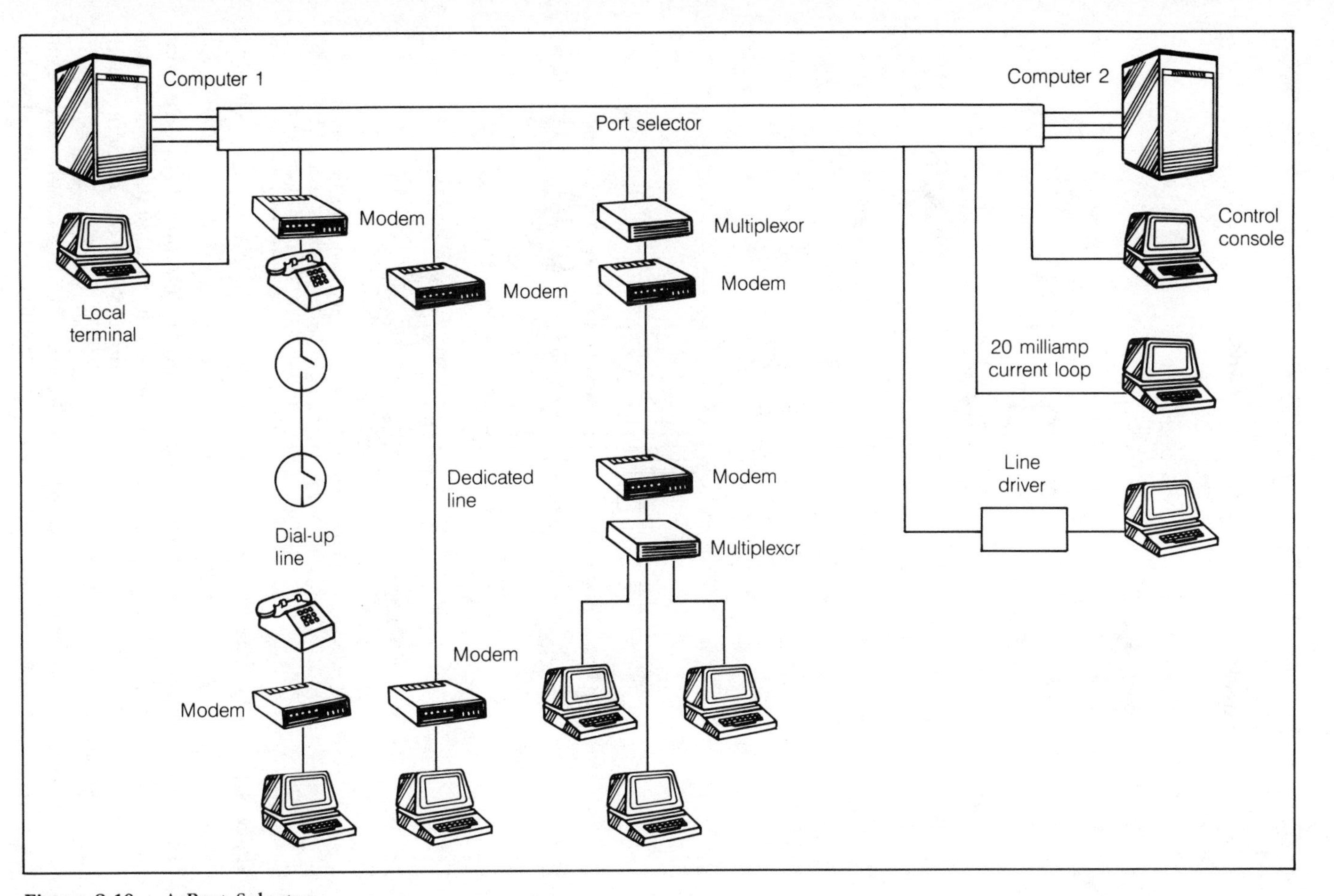

Figure 8-10. A Port Selector

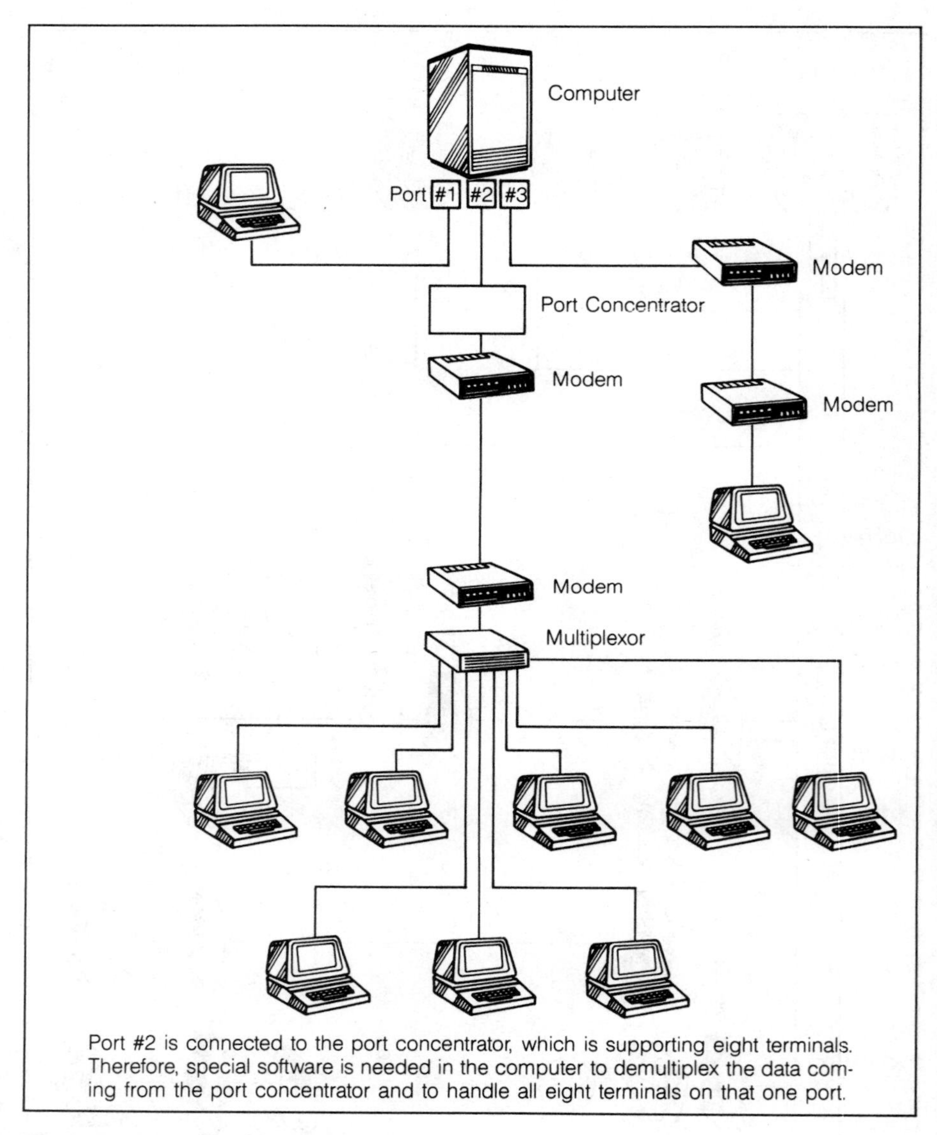

Figure 8-11. A Port Concentrator

of the stat mux, such as synchronization and error detection and correction. However, it does not break down the data; the data from all of the devices that have been multiplexed onto one line are all fed directly into one computer port (see Figure 8-11).

The other black box devices we have discussed in this chapter are "user transparent"—that is, no changes to terminals, computers, software, and so on are required to use them. However, the port concentrator does require a software change in the host computer to "demux" the data coming in from the port concentrator. Whether the additional CPU time and programming effort required is worth the port savings will depend on the individual installation. For example, if you can have a maximum of thirty-two ports on your computer and they are all in use except for one or two, and you then want to add sixteen terminals that must be on-line at all times but will have only short infrequent messages to send and receive, a port concentrator may be a good approach.

REDUCING TRANSMISSION ERRORS

As we have mentioned before, basic communications techniques between dumb terminals and their host computers were designed for transmission distances of less than fifty feet. When these same techniques are used over longer distances (for example, when they are used over telephone lines that can be subject to electrical interference from such sources as power lines or lightning and especially when they are used with higher-speed modems whose coding techniques are more sensitive to transmission impairments), transmission errors can be introduced. In other words, a 1 can be transmitted but a 0 can be received or vice versa, without being detected.

Parity checking, which is sometimes used with asynchronous transmission, will not detect many errors where more than one bit in a character has been changed. Therefore, when data is being transmitted between remote devices and the accuracy of the information is important, sophisticated error detection and correction schemes should be considered.

Black boxes are available that you can add on to each end of your data line, which utilize CRC error detection and retransmission schemes. As you may remember from our discussion in Chapter 4, CRC or cyclic redundancy checking is a very powerful error-detection scheme that, if properly used, can provide you with virtually error-free data transmission.

If you are planning to use statistical multiplexors on your data circuits, you will probably not need to add separate black boxes for error detection. Stat muxs typically include CRC error detection and correction, resulting in a virtually error-free data line as well as allowing several terminals to share one line and thus reducing line costs.

Character Delay Problems

Echoplexing, the technique discussed in Chapter 4 where the host computer echoes back each character as it is typed, was adopted because of the problems caused by transmission errors. If a terminal user types a 9 but a 0 is displayed (after being retransmitted by the host computer), the user knows an error has occurred somewhere and can retransmit the data. However, statistical multiplexors and error-detection/correction devices introduce delays in the data transmission that can be very bothersome with systems that use echoplexing.

Statistical multiplexors may buffer the data, or, as part of the error correction routines, the data may be retransmitted. Thus, there may be some noticeable delay between the time a character is typed and the time it is echoplexed back for display. Note that this same problem occurs on satellite circuits with their long propagation delays.

These character display delays are annoying and can make echoplexing an undesirable approach. However, with the virtually error-free circuit provided by CRC error detection and correction, you can afford to eliminate echoplexing from the host computer.

POINTS TO REMEMBER

Relatively inexpensive black boxes are available that let you reduce both your data communications line costs and computer port costs as well as providing you with essentially error-free data links. This is true even if the computer system you are using employs protocols that can only handle terminals considered "dumb" from a data communications standpoint.

By using one or more of these black boxes, you can create multiple channels on one physical circuit, convert async to sync transmission (allowing the use of higher-speed modems and eliminating the start and stop bits), implement advanced protocols similar to SDLC or HDLC, utilize multidrop lines, and implement CRC error detection. You can also provide for port concentration, port contention, or port selection in order to reduce the number of computer ports required to support your data communications system. Additionally, some of the same black boxes can be used to implement a data PBX, giving you flexibility and convenience in establishing communications between various devices and computers. In most cases, adding these black boxes does not involve making changes to your computer system, as most of them (with the exception of the port concentrator) are user transparent.

Advances in digital technology have made it possible for you to take advantage of microprocessor technology at a very low price, so that you can implement sophisticated approaches to reduce your data communications costs even with computer systems designed to support only "dumb terminals."

Thus, we've discussed how to use fewer lines to transmit your data with fewer errors. In the next chapter, we discuss what options you have in selecting and connecting those lines.

9

TRANSMITTING YOUR DATA—SHORT-, MEDIUM-, AND LONG-DISTANCE COMMUNICATIONS

Data transmission can occur in many different ways over many different media. The transmission can be for very short, for medium, or for long distances.

You have hundreds of choices as to how to move data from one point to another. This chapter explores many of the options available to the user. In order to allow you to focus on the options available for your particular situation, this chapter is divided into four sections based on the distances over which the data communications takes place:

1. Communications within the computer
2. Communications between a computer and its peripheral devices
3. Local distribution systems within a building or between closely situated buildings
4. Communications between two or more remote sites.

COMMUNICATIONS WITHIN THE COMPUTER ITSELF

While the need to communicate between a terminal in San Francisco and its associated computer in New York is obvious, the fact that data communications takes place within the computer itself is less likely to cross your mind. The two major situations where data communications takes place within the computer are communication between the various "cards" (or printed circuit boards) within the computer and communication between these boards and the video display. These are the two internal communications situations that are often discussed in the literature on data communications products.

Communications Between Printed Circuit Boards

If you consider a typical microcomputer system, it consists of a central processing unit or CPU board, a memory board, and a disk controller board (in some cases, one or more of these boards may be combined). It is necessary for the CPU board to talk to the memory board, to the disk controller board, and so forth. The communications highway within the computer over which these boards talk to each other is called a *bus.* The bus consists of physical connectors into which the individual boards can be plugged and of the data paths over which the boards can talk.

The following paragraphs look at some of the buses you are likely to come across in the literature.

The *S-100* bus is a popular bus used frequently in personal computers. It derives its name from the 100 parallel common communications channels it contains. The S-100 bus tends to be used with less expensive types of computers.

The *Multibus* was developed by Intel Corporation and has become one of the most important standard buses for use in industrial, military, and computer development systems. The Multibus is also widely used in business-oriented personal computers.

Other buses you are likely to come across are the *VME bus,* the *Q-bus,* and the *STD bus.*

What is important to keep in mind is that computer boards designed for a particular bus (such as the S-100) cannot be plugged into a different bus (such as the Multibus). Furthermore, just because a board and a computer are "bus compatible" (for example, they are both designed around the Multibus), it does not necessarily mean that the board will work in that computer.

Communications Between the Computer and the Video Display

Although this is probably better classified as video rather than data communications, it is such an important part of computer communications and is so often referred to in data communications literature that we thought it would be beneficial to discuss here.

The following paragraphs look at the three methods used to transmit the information to be displayed on the video screen from the computer.

Direct or Red-Green-Blue (RGB) The typical color CRT uses three "guns," one for red, one for green, and one for blue. In the direct or red-green-blue (RGB) method, the information is sent to the CRT over three channels, one for each gun. This is the most expensive approach but creates the sharpest picture and truest, most intense colors. The direct approach is also used with black-and-white displays where only one signal is necessary.

Composite A less expensive way for the computer to communicate with the video display is to combine the three pieces of video information onto one channel. The result is to create a composite signal that carries all the information. Although less expensive, this approach does not produce the high-quality picture of the direct approach.

Radio Frequency (RF) The third approach is to transmit a radio frequency (RF) signal that can be picked up by a conventional television receiver. This is low cost but yields the poorest quality of the three approaches, and it usually does not produce acceptable results for the word processing type of applications. The RF approach is used with both color and black-and-white displays.

COMMUNICATIONS BETWEEN A COMPUTER AND ITS PERIPHERAL DEVICES

Traditionally, when all of the peripheral devices were in the same room as the computer, the peripherals were connected to the computer by cables consisting of metallic circuits of twisted-pair wires or, in some cases, parallel wires. These circuits are now one of several options for connecting a computer to its peripherals. With metallic circuits, the cables have connectors at each end that plug into matching connectors on the computer and on the peripheral devices; the connector is called the *hardware interface* (defined in Chapter 3).

RS-232C

The most common data communications connector in use is designed to comply with the Electronics Industries Association's (EIA) RS-232C standard. The connector is commonly referred to as an "RS-232" or "25-pin" connector (see Figure 9-1).

Any device with a standard RS-232 twenty-five-pin receptacle can accept a cable with a standard RS-232 plug. Twenty-two of the twenty-five pins have a prescribed use, but whether to use the various pins in the connector or to leave them unconnected is an option of each computer or peripheral manufacturer.

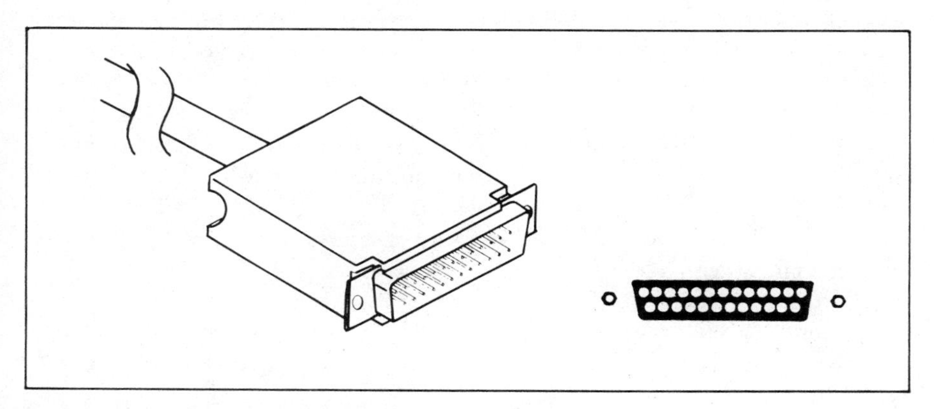

Figure 9-1. RS-232C Connector

The various pins will be connected or not depending upon whether or not the various functions assigned to those pins are going to be used by the particular device. In addition, some installations have used one or more of the three unassigned pins.

Just because two pieces of equipment can be plugged together does not mean that they can communicate over that cable. The minimum number of wires that are normally connected to an RS-232 connector is three: one for transmitting data, one for receiving data, and one signal ground or common return. Many more wires can be connected for various control, test, protective-ground, and signal functions. If a particular peripheral device is expecting one of those connections to be there and it is not, the device will not work properly.

The computer and the peripheral device also need the correct software in order to talk to each other. If a port on a computer has an RS-232 connector and software that is expecting to talk to a CRT terminal, it may be possible to use a printer plugged into the same port instead of the CRT terminal or it may not, depending upon the software in the computer and the printer.

Finally, there are two versions of the RS-232C connector with slightly different wiring, one for data communications equipment (DCE) and one for data terminal equipment (DTE). The biggest problem you will usually encounter involves pins 2 and 3. DTE sends data on pin 2 and DCE receives data on pin 2. You must therefore ensure that the RS-232C connector and cable are wired for the use intended.

If you run into incompatible devices, you can often connect them with a crossover cable that cross-connects the appropriate signals from certain pins on one connector to different pins on the other connector (such as pin 2 to pin 3 and pin 3 to pin 2). Another approach is to use a "null modem," which is a black box that reverses the transmit and receive wires and the hand-shaking signals (which involve pins 2, 3, 4, 5, 6, and 20) on the RS-232C connector but leaves the other signals alone.

Another problem you may encounter is that the RS-232C specifications call for male connectors on DTE and female connectors on DCE, but manufacturers commonly ignore this requirement. "Sex-change" adapters are available from many sources to solve this problem if you encounter it.

A word of caution is in order at this point. Although multiplexors are usually considered DCE, most manufacturers configure the RS-232C connectors on multiplexors as though they were DTE. This means that you may have problems with your cable between your terminal (DTE) and your multiplexor (DCE, but usually wired as DTE). However, your multiplexor (because it is usually wired as DTE) should connect with a standard RS-232C cable to your modem (DCE).

If you can't seem to figure out how to hook up your different RS-232C equipment, you might want to consider a breakout box (described in Chapter 13). Also see Chapter 13 for how to test an RS-232C connector to determine if it is DCE or DTE.

Standard Cables The RS-232C standard limits the length of wire or cable between two devices to fifty feet at a data rate of 20,000 bits per second or higher. In practice, however, most peripheral equipment has been designed to exceed the fifty-foot limit easily, but in most cases the distance cannot exceed 200 to 500 feet.

Extended-Distance Cables In addition to the standard cables that are designed to meet the RS-232C standards, there are special cables known as extended-distance cables that are designed to allow connections of up to 1,000 feet at slower data speeds. The field service department of the manufacturer or distributor of your computer or peripheral device can advise you as to the distance that will be practical in your situation. In many cases, the only true way to find out if a longer cable will work is to try one.

Other Connectors

In addition to the RS-232C, several other "standard" connectors and cables are available. Individual computer and computer peripheral manufacturers use the local connection medium and the connector that they think will provide the most sales. This practice, along with the need for signaling capabilities not provided with the RS-232C connector, has led to the use of a number of non–RS-232C connectors. (Some of the information in the following subsections repeats information provided in Chapter 3 because you particularly need to understand these options in order to make sense of most data communications literature.)

RS-449 The RS-449 connector and cable are designed for data rates of up to 2 million bits per second and permits cable lengths of up to 4,000 feet depending on the speed. Many computers and peripheral devices can support the RS-449 interface, which is one approach to overcoming the speed and distance limits of the RS-232C.

Current Loop This interface has been widely used in connecting teletypewriter terminals and is available on other devices as well. Although once very popular, it is being phased out. One significant advantage of this interface is the long loop (or line) lengths it can support, which may make it worth considering if you are having a loop-length problem. A number of different types of peripherals can still be ordered with a current-loop interface, and RS-232C-to-current-loop converters are available.

IEEE 488 The IEEE 488 interface was developed to connect various test instruments (such as oscilloscopes, voltmeters, and so on) so that they could be controlled by a central computer. It is often referred to as the "instrumentation interface" and is also known as the IEEE 488 bus. It is a parallel interface that is also used with such devices as printers and light pens.

Parallel Versus Serial Interfaces Many printers are available with either parallel or serial interfaces. The major consideration for most users besides whether the proper port is available on the computer will be distance. In most cases, the printer can be located much farther from the computer with a serial interface than with a parallel interface.

Using a Medium Other than Metallic Wire

In addition to metallic wire cables, peripherals in the computer room can also be connected with coaxial cables and optical fiber cables. The choice of medium in the computer room situation is usually determined by the computer manufacturer, so you are likely to have little choice in this area. However, LAN's (discussed in Chapter 10), digital PBXs (discussed later in this chapter and in Chapter 10), and data PBXs (discussed in Chapter 8) can be implemented for computer room situations if they suit your needs. LANs, coaxial cables, optical fibers, digital and data PBXs, line drivers, and modems offer ways to extend the distance between a computer and its peripherals.

COMMUNICATIONS FOR LOCAL DISTRIBUTION SYSTEMS

As more people within an organization find a need to use the computer, it becomes desirable to locate personal computers, computer terminals, and printers at various places throughout the building. Companies often expand to multiple buildings at a particular site. Sometimes all buildings are on the same piece of property; sometimes they are across a parking lot or a public street. Each of these situations presents problems in how to connect the various personal computers or computer peripherals to the main computer (or computers), which may now be across a highway or in a different building. There may even be a need to connect several computers to the same peripherals with the computers being located in different buildings.

Wiring devices together within a building or a complex of buildings requires a *local distribution system* for data communications. This distribution system can take the form of wiring each device directly to the central computer or to several distributed computers, or it can take the form of a local area network, possibly based on a PBX as discussed in Chapter 10.

Twisted-Pair Wires

The wires used for these connections may be the same twisted pairs of wires used in the RS-232 cables when the peripherals and the computer were all in the same room. Normally, twisted pairs have a bandwidth in the 100 kilohertz to two megahertz (MHz) range, and the distance that signals can be transmitted without repeaters is relatively short compared to connections using coaxial cables or fiberoptics. Signal alteration at ten megahertz on telephone-company-type twisted-pair wires is about 70 dB/km (decibels per kilometer), which is high compared to coaxial cables (10 to 20 dB/km) or optical fibers (less than 5 dB/km).

Coaxial Cable

The "wires" may also consist of coaxial cable. Coaxial cable or "coax" consists of only one wire in the center surrounded by insulation and covered by a wire mesh or foil shield (see Figure 9-2).

Coaxial cable can carry a signal that occupies a very broad frequency band, or it can carry many simultaneous data conversations; thus, although it has only one wire, it can carry much more data than the twenty-five wires that can be connected to the standard RS-232C connector. In fact, one coaxial cable can replace up to 1,500 twisted pairs of wires. It also is relatively immune to "noise" from electromagnetic radiation. These two properties have made coaxial cable an important factor in local area networks.

Typically, coax cable has a bandwidth of approximately 300 MHz and a signal attenuation at 10 MHz of only 10 to 20 dB/km.

Optical Fibers

"Wires" can also be glass wires or optical fibers. Optical fibers are very thin, flexible glass rods that can be used to carry data encoded on light signals over short or long distances. The optical fibers are normally used in conjunction with laser-generated light systems.

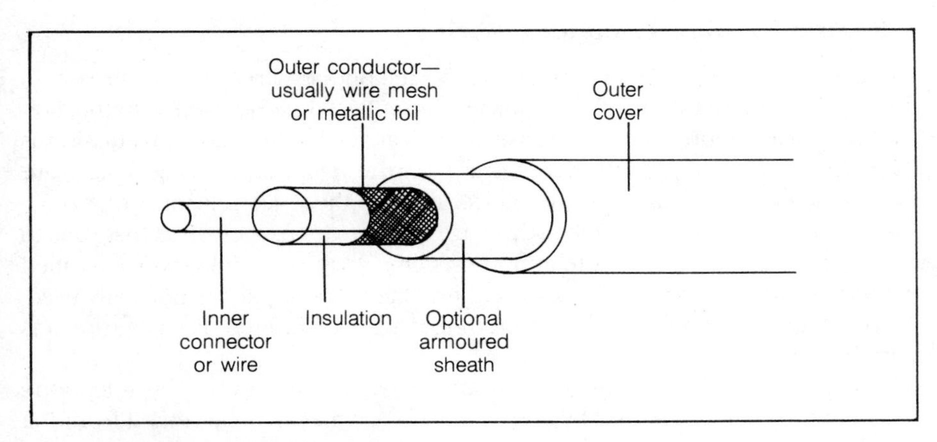

Figure 9-2. Coaxial Cable

Optical fibers have an even wider bandwidth than coaxial cable, can handle extremely high data speeds, are virtually immune from noise caused by electromagnetic radiation, and are also virtually immune from unauthorized tapping. Optical fibers can thus be an ideal choice in a very "noisy" factory (where electromagnetic interference—EMI—or radio frequency interference—RFI—is a problem) or in a high-security installation where data security is of the utmost importance. The bandwidth of optical fibers is virtually unlimited, and the signal attenuation is typically less than 5 dB/km at 10 MHz.

Using Alternating Current (AC) Power Wiring

Another and rather unusual technique is to use the 110-volt AC power wires already in place in the building to carry the data communications. This technique is referred to as the *power line carrier (PLC)* approach, since a radio frequency (RF) carrier signal is superimposed on the AC power line and the RF carrier is then modulated to transmit the data. With this AC wiring or PLC approach, a computer terminal can be moved anywhere in the building, plugged into any 110-volt AC outlet through a power-line interface unit, and will not only receive power but also will be able to communicate with the computer, all without the need for any special data communications wiring.

The major drawback of this approach has been the limited bandwidth available and its susceptibility to noise. However, advances in this technology make it worthy of serious consideration, especially when a few remote devices (such as remote metering equipment or data entry equipment) will be moved constantly from one location to another in the same building or group of buildings. For example, Data Link Corporation's 110 series PLC modems allow up to eight channels at 9600 bps and, by means of an addressing scheme, these modems allow multiple terminals to share each channel.

Utilizing Existing Telephone Wiring

Yet another wiring technique is to use the telephone wiring that is already in place. This can be done with a similar technique to the one used with the 110-volt AC wiring approach. In this case, an RF carrier for the data transmission is superimposed on top of the voice traffic on the telephone wiring. Filters are used to keep the RF carrier from interfering with the voice transmission.

Another approach is to take advantage of the extra pair of wires that runs to each telephone in many PBX telephone systems. Most older PBXs were installed with two pairs of wires per phone, yet only one of these pairs is normally used. Depending on the distances involved, line drivers or limited-distance modems may be required.

You need to be careful with these approaches, however. The telephone system wires may be owned by your company, or they may be rented from the local telephone company. In the latter case, the local telephone company may impose limitations on the use of their wires or charge additional fees for the use of the wires even if they are already in place. Also, before you use the extra telephone wires, it may be necessary to test and document the wiring; this can be labor intensive and quite costly.

A final way to utilize the existing telephone wiring for data transmission is to use a PBX specifically designed for both voice and data transmission. Two examples of digital PBXs designed for both data and voice are AT&T's System 85 and ROLM Corporation's CBX. Both of these digital PBXs provide for digital telephone sets that have CODECs built into the telephone set itself. This means that the voice signal is digitized at the telephone set and can be combined with data from, say, a CRT terminal connected to the telephone set; then both the digitized voice signal and the digitized data signal can be transmitted on the same wires.

These digital PBXs require only one or two pairs of wires from the PBX to the telephone in order to transmit both voice and data simultaneously. They can save you the cost of expensive data communications wiring such as coaxial cables.

If you are considering a new telephone system or considering purchasing the telephone system you are currently renting, you should seriously consider a PBX that would handle both voice and data.

Other Communications Media

If it is not practical to use existing wiring or to install new wiring in your building or between buildings that are close to each other, there are several other alternatives you should consider. (The following sections apply mainly to communications between buildings, but dial-up and dedicated lines can sometimes be used within a building.)

Dial-Up Circuits It is possible to use dial-up circuits either on your company's own PBX system or on the telephone company's local dial-up network to connect equipment within a building or in two closely situated buildings. In these cases, modems are used with the phone equipment just as they would be for long-distance calls. In some instances, short-haul modems will work on these dial-up circuits; with digital PBXs, no modems will be required, but some type of data termination equipment may be required.

Dedicated Circuits Sometimes dedicated circuits can be rented from the telephone company for use within your building, and they are usually available to connect two nearby buildings. It may also be possible to run your own wires, either buried underground or strung overhead, although you may have trouble getting right-of-way permission, and the cost of running your own wires may make this not feasible. Again, line drivers or limited-distance modems may be required.

Nonwired Techniques Nonwired techniques such as microwave, digital radio, infrared light, and laser transmission systems are all practical approaches for data transmission between nearby buildings if you have a clear line of sight between the two buildings. While these approaches are usually only economic in special situations, you should consider them if you have a reasonably large amount of data to send between two nearby buildings and if the cost of telephone company circuits is high or if those circuits are unavailable. Suppliers of these systems often advertise in the more popular data communications magazines and can advise you of the practicality of using their systems in your situation.

One word of caution: When using nonwired systems, you should consider carefully the effect on your business of a system outage. Outages of nonwired systems can be caused by weather, power failure, or a failure of a component in your nonwired transmission system.

One form of nonwired data communications, called a digital termination system (DTS), has been authorized by the FCC as an alternative to the local loop from your local telephone company for use with digital data transmission systems. DTS uses radio frequency transmission instead of conventional wired approaches. A number of companies are expected to provide DTS, and this may turn out to be a reliable form of nonwired local-loop data transmission with backup and service being provided by a common carrier.

COMMUNICATIONS BETWEEN TWO OR MORE REMOTE SITES

If you wish to communicate between two sites that are more than a few blocks apart, one of the most common approaches is to use the public dial-up voice network.

Dial-Up Analog Circuits

Dial-up circuits are available from your local telephone company, AT&T, and the other common carriers (see Chapter 11 for a discussion of the various common carriers).

It is also possible to utilize a private switched voice network, which your company might have, or you might arrange to have access to another company's private network.

Remember that, when using a switched voice network, modems are required for data transmissions.

Dedicated Analog Circuits

Another approach to long-distance data transmission is to use dedicated circuits. These can be either point-to-point or multidrop circuits.

Dedicated analog circuits are available from the local telephone company, AT&T, and other common carriers. Companies can also install their own long-distance wired or microwave-based communications systems, but, except in very special situations (such as those involving railroads or pipeline companies who can string wire or put up a microwave system along their rights-of-way), this is not usually feasible.

Digital Networks

In addition to the analog voice telephone lines that are available, you can also have access to various digital networks. Digital networks are available on both a dedicated-circuit and dial-up basis. AT&T's Dataphone Digital Service (DDS), which offers leased channels up to 56K bps, is an example of a dedicated circuit digital service; and AT&T's Circuit Switched Digital Capability (CSDC), which offers digital 56K bps switched channels, is an example of an all digital switched circuit service. The benefits of digital networks are that they require no modems (although often they require some type of data termination equipment), can handle faster data speeds, and normally result in fewer errors than analog networks.

CSDC CSDC offers an all-digital switched circuit service that can provide alternate voice or data service over most existing two-wire local loops to the telephone company central office; then over high-speed digital links to other central offices; and then via the two-wire local loops back to homes or offices.

CSDC provides what appears to be full duplex 56K bps service to the user by transmitting data at 144K bps half duplex and using a very fast line turnaround time. Two pieces of connecting equipment are needed to use CSDC, a terminating interface equipment (TIE) into which the telephone and data terminal equipment is connected, and a network channel terminating equipment (NCTE) which is connected to the telephone company's local loop. The TIE and NCTE are connected by a cable and an eight-wire modular jack arrangement.

In order to use CSDC the connection is established via the telephone like a voice call and then is switched into data mode. When the call is switched into the data mode CSDC is an all digital end to end service. The user should be aware that CSDC provides no end-to-end error detection or correction mechanism, leaving that task up to the user.

High-Speed Digital Communications Services AT&T's Dataphone Digital Service (DDS) is a medium-speed digital service that offers channels at speeds of up to 56K bps, but high-speed digital communications services are also available. AT&T offers a family of services known as High-Speed Switched Digital Services (HSSDS). One branch of that family is the High-Capacity Terrestrial Digital Service (HSTDS), which includes ACCUNET, a land-based service. Another branch is the High-Capacity Satellite Digital Service (HCSDS), which includes SKYNET, a satellite-based service. Both ACCUNET and SKYNET offer T1 channels.

A T1 channel is AT&T's name for a digital transmission system that carries 1.544 million bps, can be implemented on a pair of wires or on a microwave channel, and is time division multiplexed into twenty-four speech-grade channels (utilizing a PCM technique for the voice transmission), into twenty-four 56K bps data channels, or into a combination of voice and data channels. The T1 channel is part of AT&T's T-carrier digital transmission hierarchy. In addition to the T1 carrier, there is a T2 carrier, which handles 6.312 million bps and ninety-six voice channels, and a T4 carrier, which can transmit at 274 million bps.

Local Area Data Transport AT&T offers a service called Local Area Data Transport (LADT), which moves both voice and 4800 bps data simultaneously over the same telephone line. LADT was initially introduced to support Videotex, an interactive system combining video displays, text, and the ability for the user to ask for and receive specific information. Typical Videotex services are home banking, news delivery on a home television or Videotex terminal, and electronic shopping from home.

LADT is often classified as a digital service since the transmission from the central office to the other central offices or data bases is performed on a digital basis. However, the transmission over the local loop is actually analog. The network channel terminating equipment (NCTE) at the customer's premises modulates the data signals from the customer's terminal and keeping them separate from the voice signal, which stays in its standard 0 to 3300 Hz range, transmits the modulated data signal at frequencies above 4000 Hz. This allows simultaneous voice and data communications unlike CSDC. LADT, also unlike CSDC, offers end to end error checking with the customer's terminal and the central office multiplexor handling the error detection and correction.

Value Added Networks These are a type of digital network that not only transmits data but also offers code and speed conversion as well as other features.

One of the biggest classes of value added networks are *packet-switching networks*. "Packet switching" is a data communications technique where the data to be sent are broken down into short messages called *packets*. These packets, in combination with packets from other customers, are then sent to their respective destinations one at a time on any route that is available. At the destination, the individual packets coming in from possibly many different routes are reassembled into the original longer message (see Figure 9-3).

If the message can fit in its entirety into one packet, the message is referred to as a *datagram*. Datagrams are normally delivered on a "best efforts" basis. That is, the network tells the sending host computer that it has attempted to deliver the datagram, but it does not keep track of the datagram nor does it tell the sending host whether the datagram was ever received by the destination device. The issue of whether the datagram was ever received is left to other programs or levels in the network hierarchy.

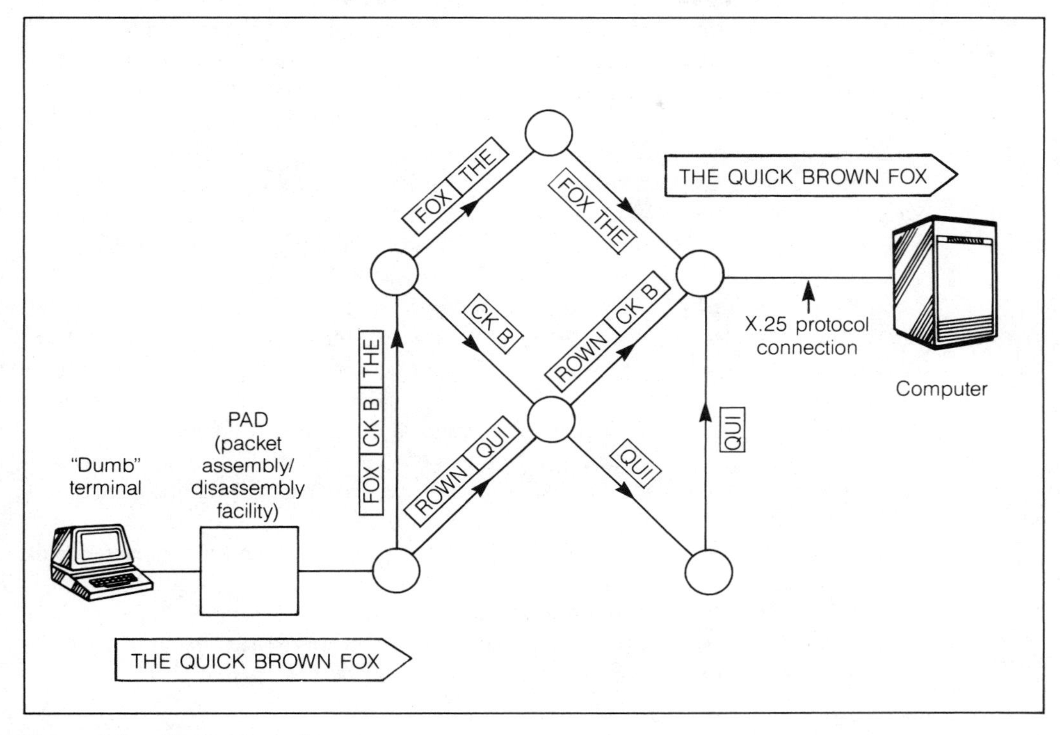

Figure 9-3. Packet Switching

Another form of communications dialogue is called a *transaction*. This is where a single packet in one direction results in the reply by a fixed number of packets from the other direction. Examples of this might be a credit-checking terminal or a door-locking system that is activated by magnetic card and is centrally controlled (such systems are often used in data processing facilities).

A more involved communications dialogue is called a *session* and occurs when a series of packets or messages are transmitted back and forth between the two end users as part of a related series of communications.

The equipment at each end of the packet-switching connection that breaks down the original message into packets and then reassembles these packets at the other end to reform the message is referred to as a *PAD*. PAD stands for *packet assembly/disassembly* facility and is a term you are likely to come across in discussions of packet-switching networks. Nonintelligent terminals that cannot interface directly to a packet-switching network with a protocol such as X.25 must be connected to the network via a PAD.

The packet-switching technique can make better use of communications circuits and thus reduce the costs of data communications because it can combine messages from many sources over the same circuits. It can also be used to interface devices that would otherwise not be compatible from a data communications standpoint, by performing code, speed, and protocol conversion. For these reasons packet switching is becoming a very popular means of data transmission.

Since they can perform these conversions, packet-switching networks can accommodate a wide variety of devices, but there must be some way for this variety of devices to interface with the packet-switching network itself. CCITT has recommended two standards for this purpose, X.25 and X.75.

X.25 is a CCITT-recommended network access protocol that includes all necessary interface standards for connecting terminals and computers (DTE) to public packet-switched networks. X.25 has become the packet network access standard in Europe and has been widely adopted in the United States as well. It specifies the electrical interface to be used as X.21 or, on an interim basis, as X.21bis, which allows RS-232C (V.24) and RS-336 (V.25) devices to be used. It also specifies a subset of ISO's HDLC as the preferred data link control protocol to connect host computers to an X.25 packet network.

At the network control level, X.25 also specifies a format for the packets and various control procedures.

X.75 is a CCITT-recommended protocol for the transference of data between two otherwise incompatible public packet-switched networks. Because X.75 is a protocol linking two networks, it is also referred to as a "gateway."

Nonwired Approaches

In addition to the "wired" approaches discussed above, the nonwired options for communications between remote sites include ultra high-frequency (UHF) and very high frequency (VHF) two-way radio, cellular radio, microwave systems, and satellite transmission. The following subsections look briefly at each of these options.

UHF/VHF Radio Companies or organizations that have access to UHF/VHF two-way radio (such as airlines, taxi companies, police departments, or companies with fleets of service trucks) can consider using those radio facilities for data communications. For example, Gandalf Data, Inc., makes the RadioModem, which interfaces an RS-232C computer port to a two-way radio.

Cellular Radio Cellular radio is becoming more widespread and available every year. It is designed to support two-way radio communication to and from mobile radios or radio telephones. These radios are normally mounted in a car or truck but are also available in briefcase models, and advancing technology is making them available in smaller sizes that can be worn on the body.

With the cellular technique, a weak radio frequency signal is used to communicate with a mobile radio while the radio is in a small geographic hexagonally shaped cell. When the radio moves from one geographic cell to the next (for instance, when the car travels a certain distance), the radio conversation is automatically handed off or transferred to another transmitter in an adjacent cell at a different frequency (see Figure 9-4). The mobile radio and the cellular base station communicate over a separate supervisory channel to coordinate the handoff automatically so that the conversation isn't interrupted.

If you are going to use cellular radio for data transmission, be careful of the data loss that will occur during the momentary signal fadeout when the mobile station is handed off from one transmitter to the next.

Microwave Digital microwave radio can be an effective and economic communications medium between two buildings that have an unobstructed line of sight between them. Microwave is generally applicable for distances of up to ten miles and is a good medium for large volumes of data or for a combination of voice and data traffic.

Microwave equipment is now available that operates reliably in the 18 to 23 GHz (gigahertz) band. However, you should remember that microwave installations are subject to both federal and local regulations. They can be affected by adverse weather conditions, and, if you do not have backup equipment, you could lose the entire high-capacity network if a critical component were to fail. Nevertheless, digital microwave radio should be seriously considered if you need high-volume voice or data transmission between two or more closely situated buildings.

Satellite Transmission Another approach, which perhaps will turn out to be the most practical of the newer data communications services, is the satellite communications system. In order to use a satellite for data communications, a message is transmitted to an earth station (which has a satellite antenna). The earth station may be in the same building as the user or may be at a remote site. The message is then transmitted by radio signal from the antenna of the earth station to the satellite, where it is retransmitted back to an earth station at the receiving site. From the receiving earth station, the message is transmitted to the end user.

The data is sent from the user to and from the earth station by conventional means, such as by local loops from the telephone company, microwave, digital radio, or digital termination systems (DTS).

Three major advantages of satellite systems for data communications are the wide bandwidth and thus the high data transmission rates available, higher-quality transmission, and their ability to provide all-digital transmission with the result that modems are not needed.

However, remember that, in order to have digital end-to-end transmission, you must set up digital local loops through such services as microwave or DTS.

A satellite transmission system introduces new problems. For example, it will take a relatively long time for the digital signal to go up to the satellite and

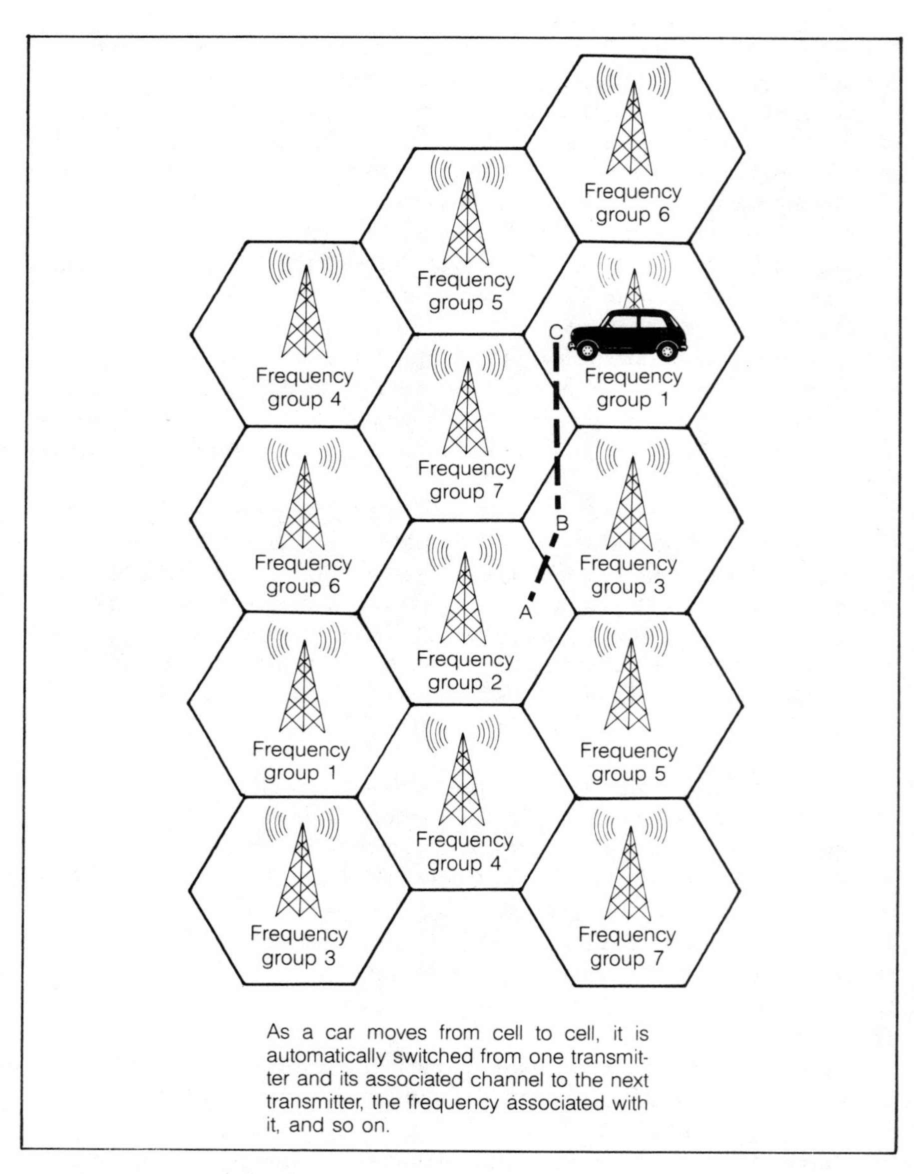

Figure 9-4. Cellular Radio

back to the earth station at the receiving end. This causes real problems with some of the data communications protocols employed, and it can make satellite circuits unusable with those protocols. Nevertheless, other data communications protocols are available that work well on satellite circuits, and satellite delay compensator units (SDCU) can be used, which electronically eliminate this problem.

Another problem is that a satellite radio signal being beamed back to earth will cover a rather wide area and can be intercepted by anyone with a satellite receiving antenna. Thus, satellite data transmissions are much more susceptible to interception than "wired" forms of transmission.

Nevertheless, if the cost of satellite transmission services comes down, you can expect satellite data transmission to become increasingly popular.

POINTS TO REMEMBER

In this chapter, we looked at data communications within the computer itself and described the computer bus. Then we looked at communications between a computer and its peripheral devices, examining interfaces such as the RS-232, RS-449, and the current-loop interface. We then looked at communications over short distances and discussed twisted-pair wires, coaxial cables, optical fibers, and power line carriers among other approaches.

Finally, we examined communications between remote sites and looked at dial-up and dedicated circuits, value added networks, packet switching, and radio, microwave, and satellite transmission techniques.

The next chapter examines one type of medium-distance communications in great detail—the local area network.

10

LOCAL AREA NETWORKS

No area of data communications seems to be growing as quickly as communications within a building or a group of buildings in a company's office complex. Local area networks or LANs have been hailed as the solution to this intrabuilding or intracomplex communications explosion.

In this chapter, we explore LANs, looking at their applications, developing a definition, describing what problems they can solve, and indicating the differences among them.

APPLICATIONS OF LOCAL AREA NETWORKS

The explosive growth in local communications is related to what is called "office automation," a phenomenon where historically manual tasks such as typing, filing, sorting, retrieving information, budgeting, forecasting, drafting, and copying are all being automated with the use of computers. Let's look at some examples of how office automation increases the need for local area networks; think about how these applications may apply to your own situation.

1. As companies make data bases in their central computers available to noncomputer professionals, the demand for on-line terminal access to this information and the associated communications requirements grows.
2. Some companies have multiple computers; for example, a company may have one for engineering, one for accounting, and one for marketing. In this case, the same users often desire access to more than one computer, requiring that the user's terminal be able to communicate with multiple computers.
3. Your company has installed word processing systems, and it seems desirable to have the word processors access such information from other company computers as accounts receivable information so that automated dunning letters can be sent out. Perhaps one department finds it would also be useful to have access to the information in another department's word processor so that retyping of common information can be minimized.

4. Another common situation is that individuals within the company start using personal computers in their jobs. However, they find it would be even more efficient if they could communicate between the various personal computers, sharing programs, data bases, and so on.

5. A company is considering adding a computerized security and energy management system and is told it will have to "wire" the building for the new system. The company is also told it needs to "wire" the building for the computer terminals it wants to install. Could both systems share the same "wiring" and save the company money?

6. Several engineers at one company were getting personal computers, and they each needed a large amount of disk storage for an occasional engineering problem they would be running that requires massive amounts of disk storage. Is there some way all of the engineers could share a large disk storage unit and conveniently have access to it whenever they needed it?

7. A company's telephone management system, word processing system, accounting system, and sales entry and tracking systems are all on separate computers. The company feels it could improve productivity if these systems were interconnected so that information could be taken from one system, processed by another, and printed out on a third. How can this be done?

8. A company has two mainframe computers, each serving 100 to 150 terminals. Each mainframe and the terminals attached to it use different types of coaxial cable for communications. A third mainframe is planned to be added with about 200 terminals. Many of the users of the new computer will want to communicate with the other two computers as well. It is not feasible to provide three terminals for each user, nor is one terminal with a simple three-way switch feasible, since all three computers use different communications systems.

In addition to solving this problem, you also want to allow the existing two mainframes to use some of the large disk storage available with the new mainframe. Is there a way to solve these problems?

All of these situations can be addressed with medium- to high-speed data communications systems designed for intrabuilding or intracomplex data communications. These systems are known as *local area networks* or *LANs*.

WHAT IS A LOCAL AREA NETWORK?

As with many other terms used in data communications, the definition of a local area network (LAN) differs somewhat depending upon who uses it. Nevertheless, most will agree that LANs have the following characteristics:

1. LANs are data communications networks for connecting various independent devices such as computers, terminals, word processors, printers, disk storage devices, and so forth. LANs should be differentiated from systems that connect various parts that form a single piece of equipment.

2. LANs, as their name implies, are for communicating in local situations, typically within a building or a complex of buildings such as on a university campus or within a manufacturing compound.

3. LANs are normally not public networks. They are usually used, owned, and controlled by one organization.

4. LANs have some kind of switching, selecting, or addressing capability that can be based on hardware or software.

5. They are generally thought of as being digital rather than analog media and thus do not use conventional modems.

6. LANs usually are designed to operate at speeds higher than the 1200 to 4800 bps full-duplex limits typically associated with dial-up telephone networks.

7. LANs often have the capability of interfacing with different types of terminals, other peripherals, and computers thus making it possible to connect different vendors' equipment on one network. However, you must keep in mind that LANs may or may not be user transparent. That is, the user may have to modify the existing data communications procedures in order to interface to a particular LAN.

8. Finally, two or more LANs can usually be connected together by gateways to provide larger or multiple data communications networks.

WHAT PROBLEMS CAN LANS SOLVE?

LANs solve the types of problems we discussed at the beginning of this chapter because:

1. LANs present a standard interface to a variety of different equipment, allowing all of the equipment to be connected to the same data communications system.

2. LANs allow data communications at medium to high speeds, making it practical to share disk storage devices and to communicate large amounts of data between two devices.

3. LANs have a switching or addressing mechanism that allows one device to specify or select which other device to communicate with.

4. LANs are being used in sufficient quantity to make prices drop, thus encouraging more widespread use of LANs.

Remember that all of the examples given at the beginning of this chapter and many more are feasible with LANs. If LANs are properly selected, they can increase the flexibility and productivity of the various office automation and data processing systems your company uses.

TYPES OF LANS

In order to select a local area network for your operation, you need to know something about the options available. LANs can be classified in many ways; I distinguish them from each other by the use of four areas:

1. Topology—ring, bus, star, and so on
2. Control—centralized versus decentralized
3. Transmission medium—broadband coax, baseband coax, twisted-pair wires, optical fibers, radio and so on
4. Access method—CSMA/CD, token passing, and so on.

Topology

Although LAN's can be built in virtually any topology, three types seem to predominate: the star, the ring, and the bus.

The Star The star topology is most often found when the LAN is based on a digital PBX such as ROLM Corporation's CBX, AT&T's System 85, or Northern Telecom's SL-1. In this case, each device on the LAN is connected to a network access unit (NAU, also possibly called a "data interface unit" or some similar term) and the NAU is connected via separate wires to the PBX, where switching and other services such as queuing and port contention (or hunting) are performed.

In modern digital PBX's, voice and data can be transmitted on the same wires, thus allowing you to use your existing telephone wiring for data communications as well.

Some authors do not include digital PBX's in their definition of LAN's but treat digital PBX's as a separate type of network.

In addition to PBX-based networks, the other most common implementation of the star configuration is the typical mini- or microcomputer with each terminal or other device connected to the central processor by its own cable.

The central control point, which is the main feature of the star configuration, results in high initial costs and potential reliability problems since, if the central control device fails, the entire network fails. Generally, a failure results in a total outage rather than a slowing of throughput, but if the failure is associated with the individual line or station control equipment, the failure may be limited only to certain devices.

On the positive side, the star topology and its implementations are well known and have a large user base. They are generally easier to maintain than distributed control systems, since all control equipment is centrally located and this usually means a lower mean time for repairs. A failure in the cable or wire to a station or in the control equipment for that station generally affects only that station and not the rest of the network. Centrally controlled systems are also easier to protect from fire and other types of damage and can be supplied with backup power more economically than other configurations. Finally, existing wiring can be used, which can make the installation of this type of LAN much less expensive than other approaches.

The Ring or Loop The ring or loop topology is the basis for many vendors' LAN's. Examples of these topologies are the IBM Series I Ring and 8100 local loop systems, Apollo Computer's Domain Network, and Prime Computer's Primenet.

The devices served by the network are connected to the ring by a network interface device sometimes called a "ring interface unit" or other similar term.

The ring or loop topology can support several access schemes including token passing, CSMA/CD (carrier sense multiple access with collision detection), slotted loop access, and polling.

The loop topology is generally thought of as a centrally controlled configuration and thus has all the advantages and disadvantages of a centrally controlled system discussed in the section on the star. However, in the case of the loop, if the cable fails to any node, it could bring the entire system down.

The ring, on the other hand, is considered a distributed control system and as such is more difficult to repair and to protect against power outages, fire, and other damage. Also, as with the loop, if one node or a cable to one node fails, the entire network could fail. Depending upon the design of the ring, partial communications may be able to continue if one or more nodes fail, but often a failure in a single node can take out the entire network.

Some of the advantages of the ring are: they are usually less expensive to get started with than a centrally controlled system; they can support very high data rates in a burst mode because each node has the full bandwidth available (a *burst mode* is in an environment where data are sent in short but infrequent bursts); and they generally assure access to the network to every node even under a heavy data load.

The Bus The bus topology is the third major topology used in LANs. Most famous of the various bus-based LANs is Ethernet. Other LANs using a bus topology are Zilog's Z-Net, Wang Laboratories' Wangnet, and Corvus Systems' Omninet.

Devices are attached to the bus by a network interface device often referred to as a *bus interface unit (BIU)*. Although Ethernet has made the CSMA/CD access method well known for the bus topology, other access methods (such as various versions of token passing) can be implemented with this topology.

The bus topology can be implemented with either centralized or decentralized control and thus has the associated advantages or disadvantages of those systems as discussed in the preceding paragraphs. Generally, the bus topology is easy to add additional stations to. Because the bus system broadcasts messages to all stations, there may be some addressing and engineering problems due to multiple pathways, and these can increase costs.

Control

LANs can also be usefully classified based on whether they have centralized or decentralized control.

Centralized Control PBXs are a prime example of centrally controlled LANs. Another includes loop-based LANs that utilize a central loop controller with a form of token passing that is really multidrop polling. Although bus-based LANs are usually thought of as having decentralized control, those bus-based LANs that utilize the multidrop polling form of token passing are normally centrally controlled.

Decentralized Control Decentralized control is the standard for CSMA/CD systems such as Ethernet as well as the approach used in most ring-based token-passing systems.

Transmission Medium

Digital PBXs utilize twisted-pair wire cables, and this same medium is used in some non-PBX LANs such as Corvus System's Omninet and Nestar Systems' Cluster/One.

Wire cable has several significant advantages over coaxial cable and other transmission media. Wire cable is generally less expensive than coax and in many cases may already be installed. PBX systems have traditionally been installed with one extra pair of wires to every telephone. In some cases, these extra wires can be used as the basis for an LAN.

On the other hand, wire cable emits and is susceptible to electromagnetic radiation and radio frequency (RF) interference. Also, for a number of practical reasons, wire cables are usually limited to much lower transmission rates than coax cables or optical fibers.

Most non-PBX-based LANs utilize coax cable, either baseband or broadband (these terms are defined later in this chapter), as the transmission medium. Coax, although generally more expensive than wire cable, is much more immune to noise and capable of much faster transmission speeds. Coax cable has been in use for many years in the cable television (CATV) industry and for data communications, and it is well supported with installation and maintenance hardware and techniques, all of which help make coax very acceptable for use in LANs.

Of course, other media, such as optical fiber and radio, can be used as well. Although optical fibers have a very wide bandwidth and are capable of much faster transmission rates than other media, they are difficult to use in LANs because of the problems of inserting taps into optical fibers to connect devices to the network.

Radio is generally not used in LANs except where radio signals are carried on the AC power wiring that runs throughout a building. Some LANs have been designed to use this technique, which shows promise for situations where a relatively small number of devices will be used and where these devices will be moved from place to place frequently. Using the existing AC power wiring can save the expense of wiring the whole building. However, these types of systems usually are restricted to slower data speeds.

Access Methods

Carrier Sense Multiple Access with Collision Detection (CSMA/CD) This is the contention access method used with Ethernet and many othr bus-based systems. CSMA/CD was discussed in detail in Chapter 6. It gives reliable throughput; however, it is generally considered more complicated to design and implement than simple token passing.

Token Passing This is an access method generally associated with rings. In its purest form, a token or message is passed from node to node in sequential order. When the receiving node gets the token, it does not rebroadcast the token until after it has broadcast its data messages. Therefore, the receiving node is the only one with the token and thus the only node allowed to broadcast. This presents a serious problem if the token gets lost, since then no node is allowed to broadcast and they all must wait. Token passing systems are therefore often designed with failure modes that include time limits after which a node is allowed to generate a token. On the other hand, this can result in multiple tokens.

Another form of access that is called token passing is, in reality, a version of polling. In this approach, a token or special message (a "go ahead and transmit" message) is sent to a specific node by including that node's address. After transmitting, the node either changes the address and sends the token out, or the central controller sends the token out with a new address. This form of token passing is usually implemented on buses since the bus topology has all nodes listening to all messages and thus needs the node address included in the token.

IEEE-802 LAN Standard

The IEEE's proposed IEEE-802 LAN standard interface defines three network access method/topology combinations:

1. CSMA/CD implemented on a bus topology
2. A token passing scheme implemented on a bus topology
3. A token passing scheme implemented on a ring topology.

It is reasonable to expect that these three access method/topology combinations will become, along with digital PBXs, the dominant LAN formats.

Front-End/Back-End LANs

Another means of classification of LANs that you may come across is front-end/back-end. These terms simply refer to whether the LAN is primarily intended to connect computers and their mass storage devices together (for example, two computers with ten disk drives and two tape drives), in which case it is called a *back-end network,* or whether it is intended to connect terminals to each other or to the host computers, in which case it is called a *front-end network.*

The primary difference is that back-end networks are intended to handle large blocks of data (like files) where response time is less important than with front-end networks. Conversely, front-end networks are designed for more frequent but shorter messages where response time is a critical factor.

Baseband Versus Broadband Transmission

Data transmission over coaxial cable can take one of two basic forms: baseband or broadband.

Baseband In baseband transmission, an electrical signal is applied to the coaxial cable, and one and only one signal is transmitted on the cable at one time. Thus, with baseband transmission, the entire bandwidth of the cable is used for a single channel. This results in a single system with uniform interfaces.

One disadvantage of this approach is that the data rate is limited (although a data rate of 50M bps has been demonstrated) so that baseband LANs cannot generally carry voice or video as well as data. However, some systems have demonstrated the ability to combine voice and data on a limited basis, and besides, most companies may have little need to carry voice, data, and video on the same coax cable.

Another disadvantage is that common low-cost cable TV amplifiers and other equipment cannot be used with baseband systems. And, finally, cable lengths and thus the size of the LAN are limited.

Broadband Broadband transmission derives a number of channels from the same coaxial cable by using frequency division multiplexing (FDM). Utilizing this approach, forty or more separate channels can be created and a total bandwidth of over 300M bps can be achieved.

Broadband transmission offers numerous advantages. Multiple channels are available, some of which can be shared, some of which can be dedicated. Some channels can be used for digital signals, while other channels are used for voice in an analog format or for video. Different networks can exist on the same cable. Cable technology and equipment, which are both reliable and low cost due to the CATV industry volume, can be used with broadband. Finally, large networks can be constructed using broadband.

On the other hand, the disadvantages include a limited bandwidth on a single channel and the more expensive interfaces and retransmission facilities that are required with broadband techniques.

Ethernet

The Ethernet system is a standard established by three companies—Xerox, Intel, and DEC. The Ethernet system is available on license to other companies, and many companies have introduced Ethernet-compatible components or devices. However, it is important to remember that all levels of the Ethernet protocol have not yet been defined, so an Ethernet-compatible device may be able to transmit a message to another Ethernet-compatible device on the network, but there is no guarantee that the receiving device will know what to do with the message or even understand it.

You should be somewhat familiar with Ethernet, so I will very briefly outline some of the aspects of this system.

- The system utilizes baseband transmission on a specially designed coaxial cable (not the same version that is standard in the CATV industry).
- The data rate is 10M bps.
- Stations or nodes on the Ethernet system can be up to 2.5 kilometers apart. However, they must be spaced in increments of 2.5 meters. This spacing requirement is critical, and thus the special cable is marked with stripes at 2.5-meter increments.
- The system is designed to handle up to 1,024 stations.
- The network architecture is a bus or a branching nonrooted tree.
- The access method is CSMA/CD. If a collision is detected, the station waits a random period of time and tries again.
- The data signals are encoded with a technique known as Manchester Code, which generates a two-level code for each bit.
- The address field is quite large—forty-eight bits for the destination device and another forty-eight bits for the transmitting device.
- The data field itself can be of variable length, up to 1,500 eight-bit bytes or 12,000 bytes.

Several companies have implemented various aspects of the Ethernet system on printed circuit boards and on integrated circuit chips. These have been made available to other manufacturers for use in designing Ethernet-based products. If these chips and boards get produced in greater quantities, we can continue to expect a decline in the cost of Ethernet interface units and thus an increase in the popularity of the system.

POINTS TO REMEMBER

You should now have a good overview of LANs, including what they are, what types of situations they can be used in, and what their various topologies, control structures, transmission media, and access methods are. I have also gone over some of the advantages and disadvantages of the various types of LANs and have detailed some of the specifics of one of the most talked about systems, Ethernet.

In order to determine what LAN, if any, is right for you, you need to analyze your own needs in terms of the categories discussed in this chapter. Then compare those needs to the specific products being offered by the vendors. The introduction that this chapter has given you to local area networks, coupled with the other information in this book, should assist you in making a sound evaluation.

The next chapter looks at common carriers and at the regulations that can affect your implementation of a local area network or other type of data communications.

11

COMMON CARRIERS AND REGULATION

In all likelihood, your plans for data communications will at one point involve the use of some facility provided by what we have historically called "the telephone company." Actually, several different "telephone companies" may be involved in providing your data communications services. The services that are available to you, and the manner in which you can obtain these telephone-company services, will probably be regulated by one or more governmental agencies. This chapter gives you some insight into what services are available from the various types of common carriers, who the regulators are, and, perhaps most importantly, how common carrier regulation affects you and your data communications needs.

COMMON CARRIERS

Companies that want to provide to the general public communications services (such as voice and data transmission) historically have had to agree to regulation by one or more governmental agencies in exchange for that privilege. These regulated companies "carry" communications transmissions on "common" or shared faciliies for the general public. Hence, these companies are referred to as *common carriers.*

The largest common carrier in the United States is AT&T. Prior to the implementation of the 1982 "consent decree modification," AT&T and its Bell System had more than 80 percent of the common carrier capacity in the U.S. Later in this chapter, we will look at other common carriers who came into being largely as a result of government deregulation, which allowed competition with AT&T. First, then, let's look at the regulators and the regulations themselves.

REGULATORS

The Federal Communications Commission (FCC) regulates broadcast, telephone, and data common carriers as well as others. The FCC has assumed jurisdiction over all international and interstate telephonic and telegraphic communications as well as over the equipment registration program. For some aspects of international communication, the U.S. State Department gets involved as well, such as representing the U.S. in the Consultative Committee on International Telegraphy and Telephony (CCITT). CCITT is a subcommittee of the International Telecommunications Union (ITU), which, in turn, is an agency of the United Nations.

All providers of regulated interstate or international communications services file tariffs with the FCC. If these tariffs are approved by the FCC, the tariffs become the contract between the carrier and the customer. You should become familiar with the specific tariffs under which you will order communications services since you will be governed by the terms of those tariffs. Keep in mind that you will be governed by the terms of the applicable tariffs even if you claim ignorance of the tariffs or even if the carrier's representative told you (even in writing) that the service would be provided under terms different than allowed by the applicable tariff.

Each state has an agency that regulates communications carriers with respect to intrastate services just as the FCC does with respect to interstate services. These state regulatory bodies are known as the state Public Service Commission (PSC) or the state Public Utilities Commission (PUC). They usually regulate a wide range of services from gas, electric, and water services to trucking and railroad companies as well as communications carriers. The state commissions approve tariffs for intrastate services that, like FCC tariffs, have the force of contracts and should be read carefully and understood by anyone ordering services under them.

LANDMARK FCC AND COURT DECISIONS

Prior to what is called the *Carterfone decision,* AT&T, the Bell System operating companies, and the other operating telephone companies generally had a tariff provision prohibiting the unauthorized attachment of any customer-owned equipment to the telephone company lines. This meant that customers had to obtain all of their telephone and data communications equipment (such as modems) from the telephone company.

The Carterfone Decision

On June 26, 1968, the FCC issued its Carterfone decision. This decision held that customers could interconnect their own equipment to the telephone company's lines as long as the equipment was "privately beneficial without being publicly detrimental"—in other words, as long as the customer-owned equipment did not present any harm to the telephone network.

This decision was the genius of the "interconnect" industry, which developed the computerized digital PBX and the 1200, 2400, and 4800 bps full-duplex modems for dial-up telephone lines. It also spawned a reaction from AT&T; AT&T began requiring protective connecting arrangements such as the data access arrangements (DAAs) in order to connect customer-owned modems to the telephone network. Finally, the Carterfone decision led to the breakup and reorganization of the Bell System and increased competition in the telecommunications industry.

Registration

On November 7, 1975, Part 68 of the FCC's Rules and Regulations was approved. This allowed manufacturers to register their data communications equipment with the FCC, provided they could meet the requirements of the FCC, which were designed to ensure that the manufacturer's equipment did not harm the telephone network. AT&T appealed this ruling, and it wasn't until 1977 that the full FCC registration program went into effect. As a result of this FCC order, all new equipment that is connected to the public telephone network after June 1, 1977 (and this includes modems as well as equipment provided by the "telephone company") must be registered by the FCC.

Registered equipment can be connected directly to the telephone network and does not have to be connected through protective connecting arrangements such as DAAs. Consequently, all communications equipment manufactured today that is designed to be connected directly to the telephone network must be registered with the FCC. Registered equipment will have both a registration number and a ringer-equivalence number printed on it. You must give both of these numbers to the telephone company before you can connect the equipment to their lines.

Execunet Decision

On June 3, 1971, when the FCC authorized companies such as MCI Telecommunications Corporation to provide private-line service in competition with common carriers like AT&T, the FCC apparently had not intended for MCI to be able to provide message toll service (MTS)—that is, the ability to use MCI's private lines to complete calls from any telephone to any other telephone by interconnecting with the local telephone company's central office. However, in October 1974, MCI offered this service under the name of "Execunet," and in July 1975 the FCC found MCI to be in violation of its tariffs and attempted to prohibit MCI from offering MTS through its Execunet service. MCI fought that decision and won in the U.S. Court of Appeals.

The FCC continued to oppose MCI, but MCI's position was upheld by the courts, and thus AT&T found itself facing true competition in the long-distance direct-dial MTS business. As a result, consumers soon had the option of several different dial telephone services at significant price savings.

1982 Modification of the 1956 Consent Decree

On January 8, 1982, the U.S. Justice Department and AT&T settled an eight-year-old antitrust suit that the Justice Department had filed against AT&T by agreeing to the breaking up of the Bell System. The proposed settlement was in the form of a modification to the 1956 Consent Decree against AT&T. (The 1956 Consent Decree resulted from a previous antitrust suit by the U.S. Government against AT&T.)

On August 24, 1982, U.S. District Judge Harold H. Greene, who was presiding over the U.S. Government's antitrust suit against AT&T, preliminarily approved a modification to the 1956 Consent Decree. This modification is referred to as the "Modified Final Judgment" or the "Greene Decision."

On August 5, 1983, Judge Greene gave his final approval to the modification but only after imposing several severe modifications of his own. The Greene Decision resulted in the restructuring of not only AT&T and the Bell System but also of the entire U.S. communications industry.

The Greene Decision included the following:

- AT&T must divest itself of its twenty-two wholly owned Bell Operating Companies (BOCs)
- The divested BOCs can market but not manufacture new customer-premises equipment such as telephones, computer terminals, and modems
- The BOCs can keep the "yellow pages" business
- AT&T must give up the use of the name "Bell" except for Bell Telephone Laboratories and international use
- AT&T's long-distance competitors receive equal access and equal facilities from the BOCs
- The provisions of the 1956 Consent Decree that restricted AT&T from entering the data processing and other markets were rescinded.

The result of this historic event was to increase substantially the competition in the communications industry and to complicate the data communications user's life. Users now have to deal with more entities in getting their end-to-end data communications established, and they also have to assume more of the data communications network responsibilities themselves.

Only time will tell whether, as a result of the 1982 modification to the 1956 Consent Decree, prices will go up or down, whether the competitors will become weaker or stronger, and whether the extra management effort required by the user is worth the change.

DIFFERENT TYPES OF REGULATED CARRIERS

The term common carrier has been used traditionally to refer to the telephone operating companies—that is, those regulated companies who supplied telephone equipment, central office switching capability, and telephone transmission facilities. These included AT&T (and its Bell System operating companies), GT&E, Continental Telephone, CENTEL, and over 1,000 smaller telephone companies operating throughout the United States.

AT&T

As a result of implementing the Greene Decision, AT&T not only divested itself of its twenty-two wholly owned local operating companies but also reorganized the remaining AT&T into a number of separate companies. Since you will undoubtedly deal with AT&T in your data communications undertakings or readings, I will give you a brief overview of the "new AT&T."

AT&T has five important subsidiaries that you should be familiar with:

1. Bell Telephone Laboratories—the research and development arm of AT&T
2. Western Electric Company—the manufacturing arm of AT&T
3. AT&T International—the AT&T company that supplies equipment and services to international markets
4. AT&T Communications—the arm that supplies switched and dedicated voice and data service (this is the company you will order your AT&T dedicated or leased data lines from)
5. AT&T Information Systems—the arm that supplies voice and data terminal equipment, such as the AT&T System 85 computerized digital PBX, and enhanced transmission services, such as AT&T's Net 1000.

International Record Carriers

The International Record Carriers (IRCs) are common carriers that provide communications service internationally. The IRCs initially provided only nonvoice services such as telex. However, as of December 8, 1982, IRCs can provide a full range of communications services including voice, data, telegraph, and telex.

Other Common Carriers

A number of common carriers were approved by the FCC after the Carterfone decision; these became known as "specialized common carriers." Originally, these specialized common carriers were only to provide private-line service. However, as a result of the Execunet decision and the Consent Decree modification, these specialized common carriers were able to offer message toll service and to interconnect fully into the local telephone operating company central office. Now they are referred to as the *other common carriers (OCC)* to differentiate them from AT&T and the other operating telephone companies that were known as the "common carriers."

The other common carriers typically provide voice or data circuits on either a private-line or switched basis over circuits either leased from a common carrier at bulk rates or on facilities built and owned by the OCC. These circuits provided by the OCC must be interconnected with the telephone operating companies' facilities at the local central office, unless the OCC has the ability to transmit directly to the customer's premises by means of radio, its own wires, or other facilities. (This is known as "bypassing" the local telephone operating company.)

The best known of the early specialized common carriers were the land-based MCI and SPCC (now called GTE-Sprint) and the satellite-based SBS.

As a result of the Greene Decision, the OCC's will be able to compete on a completely equal basis with AT&T. We can expect that many of the services provided by the OCC's will be essentially identical to those offered by AT&T but may differ in price and quality.

Value Added Carriers

The value added carriers historically did not construct their own transmission facilities but rather leased them from others. Value added carriers use standard transmission facilities (either their own or ones leased from other carriers) and then add features to provide computer-based enhanced communications services such as packet switching. These enhanced networks are referred to as value added networks (VANs). You will find that most VANs are packet-switched networks and that sometimes these two terms are used interchangeably. Two of the better known VANs are Tymnet and GTE Telenet.

Radio Common Carriers

A specialized group of common carriers, those who supply telephone service by radio link, have become much more important with the advent of cellular radio systems. Cellular radio systems use computerized switching systems to switch a customer from one low-powered radio transmitter to the next as the customer's car moves from one geographic "cell" to the next, thus greatly increasing the number of channels effectively available for mobile radio-telephone use.

These radio common carriers provide mobile radio-telephone service (that is, car- and briefcase-type radio-telephone service).

POINTS TO REMEMBER

As a data communications user, you will most likely deal with one or more of the regulated common carriers in addition to your local "telephone company." This chapter should give you enough insight and background to understand how the various carriers relate to each other and to be sure you are aware of the importance of tariffs in your dealings with regulated carriers.

The next chapter gets down to the specific decisions you need to make in order to purchase the data communications equipment you need, and it shows you how to perform a cost-benefit analysis of your expenditures.

12

SELECTING DATA COMMUNICATIONS EQUIPMENT AND CARRIERS

You have read the preceding chapters, looked at data communications literature, and made what seemed to be the difficult decisions—which type of network to use, the configuration of the network, which transmission medium to use, and how much special equipment like multiplexors and error-correction devices you need. But now, after you thought you were finished with the tough decisions, you find yourself faced with still another one, a decision that may seem even more difficult. Now you must decide on the specific pieces of equipment to select and which common carriers to use.

In this chapter, we look at the important areas you need to consider in order to make these decisions, and we provide examples of how to do the economic analysis required to make a good decision.

WHAT SPEED MODEM DO YOU NEED?

Probably one of the first decisions to be made when choosing specific equipment is what speed modems you need in the network. This is a complicated problem in network planning, but I will discuss here some of the most important elements in making this decision.

Response Time Requirements

The first criteria are: (1) what are your response time requirements? and (2) what is the volume of data being transmitted?

In a real-time environment such as a bank's electronic teller, the response times need to be short, usually two to six seconds, and the information to be transmitted is usually quite short as well, maybe ten to 100 characters.

On the other hand, in a remote batch environment, response times of hours may be sufficient, and the volume of data may be reasonably small or may be very large.

Generally speaking, the larger the volume and the shorter the acceptable response time, the higher the speed of the modem should be.

Network Configuration

In addition to response time and volume, however, you must consider the network configuration. If a data communications line has only one terminal connected to it, it can probably afford to operate at a lower speed than a line with sixteen terminals connected to it. On the other hand, if the usage of the terminals is very infrequent and very short when used (such as in a system that controls electronic door locks or burglar alarms), then no significant degradation in response time might be noticed as the additional terminals are added.

If your network uses private lines, you should also consider how you will back up your network in case a private line fails. If you plan to use dial-up lines to back up your four-wire full-duplex dedicated lines, be sure to consider modems that can handle the backup dial lines as well as the dedicated ones.

Full-duplex modems are available that will work on dial-up facilities at up to 4800 bps, but they cost more than lower-speed modems. Therefore, the cost of going from 2400 bps to 4800 bps (which includes the cost of the modems and of the backup facilities) may be more than the value of the increased throughput.

Line Noise and Other Line Problems

One might be tempted to think that, by increasing the modem speed from 4800 bps to 9600 bps, the throughput on the line and thus the response times would be increased. Unfortunately, life is not that simple. When data are transmitted at 9600 bps, the modems are more sensitive to problems on the transmission circuit. Consequently, more errors can usually be expected when transmitting at 9600 bps than at 4800 bps. If an error correction scheme such as go-back-N is used, large amounts of data may have to be retransmitted every time an error is detected. Thus, the actual throughput of clean usable data may be less with the higher-speed modem than with the lower-speed modem.

When considering high-speed modems, then, special attention should be paid to the condition of the transmission media being used and the ability of the modem to transmit and receive data with minimal errors under those line conditions. Otherwise you may find that faster is in fact slower!

Line Speed Versus CPU Speed, Disk Speed, and User Speed

The restrictions imposed by the speed of the central computer or the speed of the user may also mean that increasing the modem speed will not increase the throughput or response time.

Remember that the speed of transmission is only one component of the response time. The data must be entered into the terminal, transmitted to the computer, the computer must process the data, usually access the disk files, and then transmit it back to the user, who must read it and react to it.

If the computer processing time (including the disk access time) is fifteen seconds, and if the user is taking ten seconds to read and interpret the data, but the transmitting of the data itself is only taking two seconds, then it is easy to see that reducing the transmission time by 50 percent to one second will have little effect on the total response time.

In short, you need to look at many factors in selecting modem speeds so that the cost of the modems can be balanced against the value of the overall throughput that will result.

SELECTING EQUIPMENT

This section looks at the major issues of equipment selection: finding out what's available, deciding what features you need, establishing the equipment's availability and its reliability, looking at equipment maintenance, planning for future additions, and comparing equipment costs.

How to Find Out What Is Available

Data Communications Magazines When you start to look for the specific pieces of equipment that you need, you want first to get an overview of what is available in the marketplace. Luckily, manufacturers want to sell their products so they advertise. Among the best sources of information, then, are the various data communications magazines and, to a lessser extent, the data processing magazines.

Buyer's Guides For a reasonably complete listing of all of the products available for a specific function, such as modems or multiplexors, there are buyer's guides for data communications equipment.

Distributors and Manufacturers' Representatives Data communications equipment is normally sold through distributors or manufacturers' representatives who are listed in the yellow pages of your telephone directory. If you are not in a large city, try looking in the yellow pages of the closest major city. These telephone directories are available at your local telephone company or library. The distributors or reps will be glad to send you literature on their products and provide you with the information you need.

Distributors and manufacturers' representatives can be a particularly good source of help if you need a variety of data communications equipment from a variety of manufacturers. A strong manufacturers' rep such as MOXON Electronics in California can help you integrate equipment from different manufacturers in order to solve your communications problems.

Although manufacturers' reps are an excellent source of information and can be particularly helpful in integrating equipment from various manufacturers, you should remember that the reps generally do not carry competing products. Thus, the rep will be offering or recommending only those products that the rep carries. You may have to consider several sources if you want a broader picture of what is available.

Data Processing Equipment Suppliers If you already have data processing equipment or have narrowed the choice of a computer vendor to a few companies, ask these companies' representatives for help with your data communications choices.

Surveys There are companies, such as DataPro, that survey users and rate all types of data communications equipment. Although these surveys can be expensive and sometimes have a relatively small sample size, they can also be quite useful.

Friends, Associates, and Others Finally, you should ask friends, associates, and even strangers in the data communications world what equipment they use and why, or what equipment they recommend and why. Don't be afraid to call and ask for the manager of data communications or data processing at a company you think may have a network or communications situation that is similar to yours. Introduce yourself and ask them if they can spare a few minutes to help you with a problem you are having. If they are busy, offer to call back at a more convenient time. They are usually more than happy to tell you about their successes and how they solved their problems. Take advantage of other people's experiences; there is no need to make the same mistakes they did if you can avoid them.

Now that you've got some idea of where to look for the thousands of pieces of equipment available, what should you look for to help narrow the choice?

What Features Do You Require?

Probably the first area to consider is what features you want and whether you really need those features. Although not always true, in most cases the equipment with the most features costs more than the equipment with the least features. So make a list of features available on the various pieces of competing equipment. Then go through the list and decide which features you really need, which features would be nice to have, which features you really have no need for, and finally which features, if any, you specifically don't want. That way you can review the various offerings against the list and narrow your choice considerably.

Availability of the Equipment

In deciding which equipment to get, you need to consider its availability, both now and in the future.

Sometimes manufacturers announce equipment that can't be delivered for many months. If the equipment is available now, ask the vendor about the lead time. What happens if you need more on an emergency basis, either for replacement of a defective unit or for unexpected expansion of your network? How quickly can you get it? Where do you get the equipment from? Is it stocked locally, or does it have to be ordered from the factory back East (or out West)? How will this affect your delivery times if you need the equipment in a hurry?

Reliability

In addition to considering features and availability, you need to consider the reliability of the equipment. This is a hard area to judge. Sometimes, statistics on mean time between failures (MTBF) are available from the manufacturer, but the best approach in most cases is to ask your supplier for several references from people who are actually using the equipment. Then be sure you call the references and find out how reliable the equipment has been for them.

Maintenance Considerations

Unfortunately, all data communications equipment fails at one time or another. When it fails, how do you detect where the problem is and how do you get it fixed? As I explain in Chapter 13, maintaining your data communications system is often a much larger job than you anticipate, and it can be a frustrating experience since your business may suffer significantly if your data communications system is inoperative for any length of time.

If you select many elements of your data communications network (such as modems, multiplexors, async-to-sync convertors, and so on) from different vendors and each vendor uses a different group to repair their equipment, your maintenance problems are more complicated than if one service group is responsible for all the equipment in your network. There are several ways to handle this problem:

1. Select all or most of your equipment from the same vendor, but make sure the vendor uses the same service group to service all of the equipment. This is especially important if the service is contracted out to a third-party service organization, since different third-party organizations could be servicing different products made by the same company.
2. Select equipment (even if you get it from different vendors) that can all be serviced by the same third-party service organization. Some of the nationwide computer hardware service organizations are TRW, Indeserv, General Electric, and the Sorbus Service Division of Management Assistance Incorporated (MAI).
3. Plan to maintain most of the data communications equipment like modems and multiplexors yourself by stocking replacements in-house. This approach is discussed in Chapter 13.

4. Order your equipment from multiple sources but work with these sources to develop very detailed and comprehensive written troubleshooting and maintenance procedures that address the problems of coordination among the various vendors and that determine who is responsible for fixing a failure when it occurs.

5. Select equipment with good self-diagnostic capabilities. Self-diagnostic systems are usually easy to operate and can find most problems quickly.

Modularity, Compatibility, and Expandability

If you plan to use multiple modems, multiplexors, terminals, and so on at any one site, you should give strong consideration to equipment that is modular. Most modular equipment (like a modem) consists of a rack of card slots called a *card cage,* a power supply, and common control and diagnostic circuits. You can then plug in modem cards, auto-dialer cards, redundant power supply cards and so on, and these can be easily removed for repair or replacement or can be easily added if new circuits are required.

When selecting the different pieces of equipment, remember that they will have to work together. Be sure to check with your various suppliers to ensure that all of the equipment is compatible. Compatibility is not guaranteed even if all components are made by the same manufacturer.

Keep in mind that your network will probably grow over time. Consider what you will have to do to the equipment you select in order to expand the network. Will you have to throw away the equipment you are buying now? Can the equipment be expanded at a reasonable cost if necessary?

Equipment Costs

Finally, and it should be the last thing you consider, you need to look at equipment costs. Generally, the difference in cost among the various pieces of data communications hardware you are considering will be small in relation to the overall cost of the network on a monthly basis or to the cost to your company if the network fails for a long period of time.

So, only after you have decided on the equipment that you really want for the reasons discussed above should you give consideration to minor cost differences. Of course, you have to be practical—don't even consider a centralized, fully automated real-time network control system unless you feel the incremental cost of such a system would be justified in your application. Such systems are appropriate for large operations like airlines, banks, and utilities companies.

CHOOSING A COMMON CARRIER

Before the advent of competition, it used to be very easy to decide whom to order your data line from. You just called your local telephone company and they handled the whole thing; in fact, they even supplied the modems (data sets).

Then along came competition and you had to decide between the Bell System and a specialized common carrier (now called "other common carriers") like MCI or Southern Pacific Communications Company (now called GTE-Sprint). Then the 1982 Greene Decision broke up AT&T and life became even more complicated.

Your Local Telephone Company

The local loop from your office to the telephone company's office and private lines across the block, across town, or anywhere in your telephone company's local service area will all be ordered through and supplied by your local telephone company. In most cases this will be your only practical option, although direct satellite, microwave, digital radio, infrared, and a few other options do exist as alternatives to the local telephone company circuits.

Service Beyond Your Telephone Company's Local Area

For private lines or switched service that goes beyond your telephone company's local service area, you have a choice between AT&T and the other common carriers. Prices, the quality of the lines, and the quality of the service provided to keep the lines up to specifications and to repair them when necessary will differ from carrier to carrier. Your choices and options will be greater as a result of competition in this area, but the decisions you have to make will be more difficult.

AT&T The largest long-distance common carrier is clearly AT&T. After the AT&T split-up is complete, you will no longer obtain private lines supplied by AT&T just by calling your local telephone company. You will have to decide specifically to use AT&T and order AT&T service if that is what you want.

Other Common Carriers You can expect the number of common carriers offering data service and the different services they offer to continue to grow over the years. You should compare prices and especially the quality of service carefully before choosing any common carrier.

Since the various other common carriers will not all cover every city, you may end up using several common carriers to take care of your data communications line requirements. If you do, you should be sure that each circuit is carefully documented (as recommended in Chapter 13) so that repairs can be made quickly when necessary.

Cost Savings and Circuit Quality

The monthly cost of a dedicated circuit or the per-minute cost of a dialed call on the switched network of an OCC are generally lower than comparable service from AT&T. It is these lower costs that have attracted so many customers away from AT&T to the OCC's. Nevertheless, many OCC customers feel that the quality of circuits that they get from the OCC is not as good as that provided by AT&T. Evaluate the quality of circuits available in your area carefully when considering an OCC. Also, many OCC's provide voice-grade circuits that have a narrower bandwidth than those provided by AT&T, and many full-band modems will not work on those circuits. Special data-quality circuits must be ordered from these OCC's in such cases, usually at an extra cost.

Call Timing with an OCC

Another area to consider when using an OCC is the timing of the call. Some OCC's will time a call in smaller increments than AT&T, thus (on the surface) saving you money, especially on shorter calls. For example, on a call that lasts 1.6 minutes, AT&T would charge you for two minutes. Some OCCs would charge you for only 1.6 minutes.

Answer Supervision

On the other hand, if the OCC does not receive answer supervision signals from the local telephone company, then the OCC must estimate when the call starts (that is, when it is answered) and when it ends (that is, when one of the parties hangs up). To estimate when a call is answered, the OCC assumes an average ringing time and subtracts that average ringing time from the total connect time to establish a total "talking time" or "billing time." If your calls ring for an unusually long time before being answered, the OCC will still only subtract the average ringing time and thus charge you too much for your call.

Charges for Noncompleted Calls

Since the OCC doesn't know if the call was answered, they assume a very short call is a noncompleted call, and any call over that minimum length is a completed call. Thus, if your call rings for a long time and is never answered, the OCC that does not receive answer supervision will bill you for that noncompleted call.

You should therefore review your OCC phone bill carefully for any short calls, and if they were actually noncompleted calls ask for a credit.

Effect of the Greene Decision

When the Greene Decision is fully implemented, the OCCs will receive answer supervision from the local telephone company or have the local telephone company handle their billing, so you will be billed for only completed calls and the call lengths will not have to be estimated.

Satellite Circuits

Some of the common carriers will supply satellite circuits for your data communications circuits. Satellite circuits can be very good data communications circuits; however, the signal has to travel from your computer or terminal all the way to the satellite and back, and this means a propagation delay that is longer than on normal terrestrial circuits. Be sure your data communications equipment and protocols are designed to work with satellite circuits before ordering them, or you may find that your network won't work or that you experience significant delays in your data transmission.

PERFORMING AN ECONOMIC ANALYSIS

You've narrowed down the list of specific equipment you want to buy to two or three pieces of each type of equipment. However, the different vendors offer such different payment plans you don't know how to compare the cost of the various offerings.

One vendor wants to rent you the equipment for a monthly fee that includes maintenance, but you do have to pay an installation charge.

Another vendor wants to sell you the equipment for a one-time fee, and then have you contract with a third-party service company for maintenance for a monthly fee.

A third vendor will lease you the equipment for five years and will include the installation charge in the monthly lease payment. This vendor suggests that you lease a spare unit and do the maintenance yourself on a replacement basis. How do you compare the different vendors' proposals?

Including All the Costs

The first step in evaluating the costs of different proposals is to be sure you have included all of the costs. The following paragraphs point out some of the costs you should be sure to consider.

Facility Costs These include such items as: the room that the data communications equipment will be in; power and air-conditioning costs; the cost of any special power circuits or AC line conditioning needed by the data communications equipment; and the fire protection equipment (do any sprinkler heads have to be changed?)—remember, data communications equipment doesn't like to get wet.

Installation Costs These include freight, delivery, set-up and checkout of the equipment. Does the vendor pay for these, or do you?

Taxes Sales tax should be obvious, but don't forget property tax if you own the equipment. Also, if you own the equipment, you should see if you are eligible for the investment tax credit. If you are, this could reduce the cost of the equipment by up to 10 percent. Finally, although we won't consider it in our example, you might want to consider the tax advantages of accelerated depreciation.

Insurance Are you going to insure the data communications equipment, or is the vendor?

Maintenance Does the vendor include maintenance? Is it provided under contract by a third party? Do you do it yourself with extra spares you buy?

Of course, the cost of the equipment itself and the cost of financing the equipment must also be included.

How to Compare Costs

Now that you have put together a list of all of the costs for each of the vendors you are considering, how do you compare the different vendors? Which vendor's plan will cost you less? How do you compare the different payment plans when some have one-time up-front payments and some don't or when the monthly payment amounts and up-front amounts differ?

One answer is a tool of economic analysis known as *present value*. It is a technique that can be used to compare payment plans that include payments of different amounts at different times. This technique takes the payments that occur over different time periods and converts them to a uniform basis so you can compare them.

The present value approach takes all of the payments required under a specific plan and calculates their present value. That is, it calculates what the financial equivalent of the one-time up-front payment would be if all the various payments for that particular plan were made at once.

Let's look at an example. One plan calls for a $1,000 installation fee up front and a $200-a-month rental, payable each month in advance. We will calculate the present value or equivalent one-time cost of this plan over a five-year period assuming a 1 percent monthly interest rate.

The present value tables are normally arranged for payments made at the end of the month. If a payment is made at the beginning of a month, then an adjustment must be made in order to use the tables. A payment at the beginning of month 2 can be thought of as the same as a payment made at the end of month 1. A payment made at the beginning of month 1 is the same as a payment made at the end of month 0, or "up front."

The present value or the up-front equivalent cost of the $1,000 installation fee is, of course, $1,000, since it is to be paid at the beginning of month 1 (or at the end of month 0). The present value of the first month's $200 payment is $200 since it also is paid up front or at month 0. However, there are then fifty-nine payments of $200 a month required for the balance of the sixty-month or five-year period. These payments are to be made at the beginning of months 2 through 60 (or at the end of months 1 through 59).

The present value of these fifty-nine payments is found by multiplying the "present value of an annuity of 1 (PVA1) factor" for 1 percent for fifty-nine months (PVA1 – 1% – 59) times $200. PVA1 is a factor that calculates the present value of a series of equal payments of $1 each made at the end of each month for the number of months specified. The factor in this case is 44.4046 (a table of present value factors is shown as Figure 12-1). So the present value of the fifty-nine $200-a-month payments is $8,880.92. Therefore, the total present value is $1,000 + $200 + $8,880.92 or $10,080.92—or, after rounding to the nearest whole dollar, $10,081.

Another way of looking at the concept of present value is, if you put $10,081 in a bank that paid interest at the rate of 1 percent per month, then the $10,081 plus the interest it would earn would be exactly enough to pay the $1,000 installation charge and the sixty $200-a-month rental payments.

We need to look at one more present value concept before we can use this technique to compare our three vendors' proposals. What do we do if a one-time payment of $1,000 is required at the end of the sixty months in addition to all of the other payments? How do we calculate its present value? The answer is that we simply multiply the $1,000 times the "present value factor" (PV—note that this is a different factor than the "present value of an annuity of 1" factor we used before) for 1 percent and sixty months (PV 1% – 60), which is 0.5504. This tells us that the $1,000 payment at the end of the sixty months has a present value of $550.40, which we then add to the $10,081 for a total present value of $10,631.

Note that the "present value factor" calculates the present value of a single payment made at the end of the specified number of months.

Now let us do the calculations to compare the offers from the three vendors discussed above. Since we are trying to compare the three offers, we will ignore any costs that are the same, such as facilities costs.

First, we list the specific costs proposed by the three vendors.

1. Vendor One:

- Installation cost: $1,000
- Rental payment due at time of installation: $200
- Rental payments due at the beginning of each month, starting second month: $200 per month
- Maintenance fee: Paid by vendor
- Property taxes and insurance: Paid by vendor
- Investment tax credit: Retained by vendor
- Payment due at end of sixty months: None

Table 12.1 Economic Analysis Factor Tables

Months	1% per Month			1.25% per Month		
	PV Factor	PVA1 Factor	CRF Factor	PV Factor	PVA1 Factor	CRF Factor
1	0.99010	0.99010	1.01000	0.98765	0.98765	1.01250
2	0.98030	1.97040	0.50751	0.97546	1.96312	0.50939
3	0.97059	2.94099	0.34002	0.96342	2.92653	0.34170
4	0.96098	3.90197	0.25628	0.95152	3.87806	0.25786
5	0.95147	4.85343	0.20604	0.93978	4.81784	0.20756
6	0.94205	5.79548	0.17255	0.92817	5.74601	0.17403
7	0.93272	6.72819	0.14863	0.91672	6.66273	0.15009
8	0.92348	7.65168	0.13069	0.90540	7.56812	0.13213
9	0.91434	8.56602	0.11674	0.89422	8.46234	0.11817
10	0.90529	9.47130	0.10558	0.88318	9.34553	0.10700
11	0.89632	10.36763	0.09645	0.87228	10.21780	0.09787
12	0.88745	11.25508	0.08885	0.86151	11.07931	0.09026
13	0.87866	12.13374	0.08241	0.85087	11.93018	0.08382
14	0.86996	13.00370	0.07690	0.84037	12.77055	0.07831
15	0.86135	13.86505	0.07212	0.82999	13.60055	0.07353
16	0.85282	14.71787	0.06794	0.81975	14.42029	0.06935
17	0.84438	15.56225	0.06426	0.80963	15.22992	0.06566
18	0.83602	16.39827	0.06098	0.79963	16.02955	0.06238
19	0.82774	17.22601	0.05805	0.78976	16.81931	0.05946
20	0.81954	18.04555	0.05542	0.78001	17.59932	0.05682
21	0.81143	18.85698	0.05303	0.77038	18.36969	0.05444
22	0.80340	19.66038	0.05086	0.76087	19.13056	0.05227
23	0.79544	20.45582	0.04889	0.75147	19.88204	0.05030
24	0.78757	21.24339	0.04707	0.74220	20.62423	0.04849
25	0.77977	22.02316	0.04541	0.73303	21.35727	0.04682
26	0.77205	22.79520	0.04387	0.72398	22.08125	0.04529
27	0.76440	23.55961	0.04245	0.71505	22.79630	0.04387
28	0.75684	24.31644	0.04112	0.70622	23.50252	0.04255
29	0.74934	25.06579	0.03990	0.69750	24.20002	0.04132
30	0.74192	25.80771	0.03875	0.68889	24.88891	0.04018
31	0.73458	26.54229	0.03768	0.68038	25.56929	0.03911
32	0.72730	27.26959	0.03667	0.67198	26.24127	0.03811
33	0.72010	27.98969	0.03573	0.66369	26.90496	0.03717
34	0.71297	28.70267	0.03484	0.65549	27.56046	0.03628
35	0.70591	29.40858	0.03400	0.64740	28.20786	0.03545
36	0.69892	30.10751	0.03321	0.63941	28.84727	0.03467
37	0.69200	30.79951	0.03247	0.63152	29.47878	0.03392
38	0.68515	31.48466	0.03176	0.62372	30.10250	0.03322
39	0.67837	32.16303	0.03109	0.61602	30.71852	0.03255
40	0.67165	32.83469	0.03046	0.60841	31.32693	0.03192
41	0.66500	33.49969	0.02985	0.60090	31.92784	0.03132
42	0.65842	34.15811	0.02928	0.59348	32.52132	0.03075
43	0.65190	34.81001	0.02873	0.58616	33.10748	0.03029
44	0.64545	35.45545	0.02820	0.57892	33.68640	0.02962
45	0.63905	36.09451	0.02771	0.57177	34.25817	0.02919
46	0.63273	36.72724	0.02723	0.56471	34.82288	0.02892
47	0.62646	37.35370	0.02678	0.55774	35.38062	0.02826
48	0.62026	37.97396	0.02633	0.55086	35.93148	0.02783

Table 12.1 *(continued)*

Months	1% per Month			1.25% per Month		
	PV Factor	PVA1 Factor	CRF Factor	PV Factor	PVA1 Factor	CRF Factor
49	0.61412	38.58808	0.02591	0.54406	36.47554	0.02741
50	0.60804	39.19612	0.02551	0.53734	37.01288	0.02702
51	0.60202	39.79814	0.02513	0.53071	37.54358	0.02664
52	0.59606	40.39419	0.02476	0.52415	38.06773	0.02627
53	0.59016	40.98435	0.02440	0.51768	38.58542	0.02592
54	0.58431	41.56866	0.02406	0.51129	39.09671	0.02558
55	0.57853	42.14719	0.02373	0.50498	39.60169	0.02525
56	0.57280	42.71999	0.02341	0.49874	40.10043	0.02494
57	0.56713	43.28712	0.02310	0.49259	40.59302	0.02463
58	0.56151	43.84863	0.02281	0.48651	41.07952	0.02434
59	0.55595	44.40459	0.02252	0.48050	41.56002	0.02406
60	0.55045	44.95504	0.02224	0.47457	42.03459	0.02379

- Installation cost: None
- Purchase cost: $8,000
- Maintenance fee: None first year, $40 per month thereafter payable at the end of the month
- Property taxes and insurance: $9.50 per month payable at the end of the month
- Investment tax credit: $800 (10 percent)
- Payment due at end of sixty months: None

3. Vendor Three

- Installation cost: None
- Monthly fee due at time of installation: $200
- Monthly fee due at the beginning of each month, starting second month: $200
- Maintenance fee: None—price includes spares but user must replace defective unit and send spare to factory for repairs (user's time is calculated at no cost)
- Property tax and insurance: Paid by vendor
- Investment tax credit: Retained by vendor
- Payment due at end of sixty months: $900 if user wants to own equipment

Now let's look at the present value calculations for each case.

Vendor One

Payments made at the end of month 0 (beginning of month 1):

$1,000 installation fee
$ 200 monthly rent fee
$1,200 × (PV factor − 1% − 0) 1.0000 = $1,200

Payments made at the end of months 1–59 (beginning of months 2–60):
$200 monthly rent fee
$200 × (PVA1 factor – 1% – 59) 44.4046 = $8,880.92 or $8,881
Total present value = $1,200 + $8,881 = $10,081

Vendor Two

Payments made at the end of month 0 (beginning of month 1):
$8,000 purchase price
– $ 800 investment tax credit (can be subtracted immediately from estimated tax payments if you qualify for the credit)
$7,200 × (PV factor 1% – 0) 1.0000 = $7,200
Payments made at the end of months 1–60:
$9.50 property tax and insurance
$9.50 × (PVA1 factor 1% – 60) 44.9550 = $427.07 or $427
Payments made at the end of months 13–60:
$40 maintenance (no maintenance months 1–12 as equipment is under warranty)
$40 × (PVA1 factor 1% – 48) to calculate the present value of the forty-eight payments at the beginning of month 13 (end of month 12) 37.9740 = $1,518.96
$1,518.96 × (PV factor 1% – 12) to calculate the present value of the forty-eight payments at the beginning of month 1) 0.88745 = $1,348.00 or $1,348
Total present value = $7,200 + $427 + $1,348 = $8,975

Vendor Three

Payments month 0:
$200 first month's lease payment in advance
$200 × (PV factor – 1% – 0) 1.0000 = $200
Payments made at the end of months 1–59 (beginning of months 2–60):
$200 monthly lease fee paid each month in advance
$200 × (PVA1 factor – 1% – 59) 44.4046 = $8,880.92 or $8,881

Payments at end of lease (month 60):
$900 if customer wishes to buy system
$900 × (PV factor 1% – 60) 0.5504 = $495.36 or $495
Total present value = $200 + $8,881 + $495 = $9,576

For those of you who would feel more comfortable comparing monthly amounts rather than lump-sum amounts, it is easy to convert a lump-sum present value amount into its equivalent monthly amount by just multiplying the lump-sum amount times the appropriate capital recovery factor (CRF).

For example, the monthly amount, at 1 percent interest a month for sixty months, that is equal to a present value lump sum of $10,000 is $10,000 × (CRF 1% – 60) 0.02224 = $222.40.

Now let's look at all three vendors' proposals on both a present value lump-sum and equivalent monthly amount basis.

VENDOR NUMBER	PRESENT VALUE LUMP SUM	EQUIVALENT MONTHLY AMOUNT
One	$10,081 × 0.02224	= 224.20
Two	$ 8,975 × 0.02224	= $199.60
Three	$ 9,576 × 0.02224	= $212.97

You can now clearly compare the cost of the three proposals and determine in your own mind with your own set of criteria whether the differences in features or any other aspect of the offers are worth the difference in cost.

POINTS TO REMEMBER

In this chapter we reviewed some of the important aspects of selecting data communications equipment and carriers.

We first examined what you should consider in selecting the modem speed you will use. Then we talked about where you can find out what equipment is available, how you determine which features you want among all the features offered by competing suppliers, and why availability, reliability, and maintenance are important issues.

Then we discussed some of the differences between AT&T and the other common carriers, and finally we looked at how to compare the costs of competing equipment and services when they will be paid for in different ways or at different times.

In the final chapter, we cover the major issues involved in keeping your data communications network running efficiently once it is installed.

13

MANAGING A DATA COMMUNICATIONS NETWORK

Managing a data communications network can become a real problem for companies without a data communications staff. Networks are mysterious and technical; jargon, operational theory, and testing techniques can all be incomprehensible to the average uninitiated company employee. Networks break down and when they do, data communications functions stop. These functions, when they stop, often affect many aspects of a company's operations in a very damaging way. Therefore, in this chapter we will review some basic techniques that can be used to make the job of managing a data communications network easier.

When a company installs its first mini- or microcomputer system, they often give the accounting department responsibility for running the computer. Perhaps a data processing manager is hired. This doesn't present a big problem with modern small computers; they tend to be easy to install and run on a routine basis. When they fail you, call your friendly computer maintenance person. However, this simplicity rarely extends to the data communications network itself.

WHEN PARTS OF THE NETWORK COME FROM DIFFERENT VENDORS

The network is likely to consist of:

1. The lines from two or more vendors, the local loops from the telephone company, and the long-haul lines from AT&T or one of the other common carriers such as MCI
2. Modems from one or more manufacturers
3. Add-on black boxes such as error controllers or statistical multiplexors from another manufacturer
4. The computer itself

5. The terminals, which may be from one or more manufacturers (even if they are all from the same manufacturer, they may or may not be under maintenance contract or may be serviced by different service organizations)
6. And, in some organizations, the network may include voice, video, and FAX transmissions with their attendant equipment as well as data transmissions.

Because the network consists of many components with a variety of organizations responsible for the maintenance of the different components, how do you determine where the problem is? How can you be sure that the problem gets fixed quickly?

Whom Do You Call When a Problem Develops?

Whom do you call when a user at a remote site claims his or her terminal won't work? First, you have to ask yourself all of the following questions:

- Is it the terminal?
- Is it the modem at the remote end?
- Is it the long-haul phone lines?
- Is it the local loop?
- Is it the modem at the computer site?
- Is it one of the black boxes being used?
- Is it the computer (perhaps one of its communications controller boards)?
- Is it really even a problem with the network?
- Is it the application program?
- Is it the user who doesn't know how to use the terminal or how to run that program?

In other words, you have to find out where the problem is so that you can figure out whom to call in order to fix it.

Establishing and Maintaining Data Communications Procedures

In my opinion, one of the most important aspects in managing the network troubleshooting monster (and the key to answering the questions listed in the preceding section) is to develop and maintain written data communications procedures. I suggest these be written in the same format as the company's overall policy and procedures manual, if one exists. The data communications procedures manual should contain as much information on your data communications network as possible so that the information necessary to keep your system operating can be found quickly.

The data communications procedures manual should have a standard way of showing revision numbers at the top of each page (for example, a page might be headed by "Procedure #115, Page #5, REV #3—April 5, 1984"). The manual should also include a listing that shows the latest revision number for each page, so the user can be sure that the copy of the manual being used is completely up to date.

The manual should be revised whenever lines or equipment are added or deleted. It should be updated to document any quirks of your system as they are discovered and to describe the troubleshooting approaches that are developed to address those specific problems.

Background Information The data communications procedures manual is the best place to keep background information for each point-to-point circuit, each segment of a multidrop circuit, and each piece of data communications equipment.

A standard format should be developed for your manual that will work best for your organization. The manual should contain the important information you will need in troubleshooting. For example, the following information is desirable for point-to-point data circuits:

1. Name and address of the user at each end of the circuit
2. Name, address, telephone number, and contact of the carrier supplying the circuit
3. Circuit type and circuit number
4. The number of the test or trouble center and the names of individuals there that you usually work with
5. What DTE is connected to each end of the circuit and who maintains this equipment
6. What type and speed of modems are connected to the circuit and who maintains them.

Trouble Log On the back or bottom of the sheet containing background information for each circuit should be a trouble sheet or test log. (Of course, the log can also be on a separate sheet next to the background sheet.) In small installations, I like to keep this log as part of the data communications procedures manual so everything is in one place. In large installations, these logs and background sheets will probably be kept in a separate file.

The trouble/test log is used to record every problem encountered with a circuit, to whom it was reported, at what time the trouble was reported and by whom, when it was cleared and who tested and accepted the line when it was fixed. If you have test equipment, the log should also be used to record your periodic measurements on the circuit so you can see if any deterioration is taking place in circuit quality.

The log is an invaluable aid in tracking down problem circuits. It is also the key to remind you to tell the test technician about those quirks or unusual things that seem to go wrong with that circuit, since history has a habit of repeating itself. (The telephone company's test technician uses a device called a *test board* to tap into a circuit and determine where the problem is.)

If you want to reduce your network's "down" time:

1. Set up a trouble/test log.
2. Maintain the log.
3. Read the log.
4. Learn from the log.

Isolating the Problem and Reporting It Correctly One of the biggest problems in getting a faulty network operating again is isolating the problem and reporting it correctly. It does no good to report a line as being out of service to the telephone company if the problem is with a modem you are maintaining yourself. The telephone company is likely to come back the next day with "no trouble found," and you've lost twenty-four hours.

When it is a circuit problem, we have found that the people at the test board often work on the wrong circuit. Either the circuit number of the defective circuit was reported incorrectly, written down wrong, or misunderstood. Doublecheck the circuit number of the faulty circuit and have the repair person, preferably on the test board, verify the circuit number to you.

Later in this chapter, we will go through some basic test procedures that can be used to isolate where the problem is, but for now, remember that, when dealing with multiple vendors on a network, it is critical you report the problem to the correct vendor. Also, be sure you designate the symptoms as precisely as possible and do what you can to pinpoint the specific segment or piece of equipment.

Escalating a Service Request Nothing is more frustrating than reporting a bad circuit to the carrier's test center and then, two days later, still be answering those phone calls from users asking when the network will be working again, without having heard one word from the data circuit supplier.

The solution to this problem is "judicious escalation"—that is, moving the request for service up the management hierarchy of the data circuit supplier until the problem is solved.

The first step in utilizing escalation as a management tool is asking for and getting a commitment to a reporting or completion time. When you report the trouble, ask when you can expect the trouble to be cleared or when you can expect to hear back from the test center. Write down in your log at what time you reported the trouble, to whom you talked, and by what time they said they would clear the trouble or get back to you (the *commit time*).

If you have not heard by the commit time, follow up—call back—find out what's happening! A word of caution, though: The commit time may be unreasonably long, such as forty-eight hours on a critical circuit. Then you have to become more creative. Perhaps you can call the test center directly and offer to help test the line. This approach worked particularly well for us with one of the common carriers from whom we were getting most of our data circuits. We got to know all the technicians at the test center, were always very friendly and polite to them, and always asked if there was anything we could do at our end to make it easier for them. We then followed up at regular intervals. Needless to say, our data lines were always fixed quickly.

If you can't get to the test board (the phrase stands for the supplier's test technicians as well as for the device they use) directly, set your own reasonable "commit time"—say, eight hours. Then if you've not heard anything, escalate: call the supervisor and point out that you have had eight hours of down time on a critical circuit and that you can't find out what's happening. Would the supervisor check and call you back in, say, one hour (get that commit time!)?

When the commit time has passed, call the next level supervisor, explain the situation, ask for help, and ask for a reasonable commit time for the supervisor to get back to you. Always be polite, no matter how mad you are or how incompetent you think the test crew is. Try to communicate to the supervisor that you understand the pressure and problems the supervisor is facing. But point out that you really need the supervisor's help and would appreciate whatever can be done.

The escalation process can be continued if necessary.

Remember, your attitude is extremely important. Someone's job ticket has to be at the bottom of the pile of repair requests, and if you get the technicians or supervisors mad at you, it may be yours.

TOOLS AND TECHNIQUES TO HELP YOU ISOLATE PROBLEMS

Before you can report a problem to the vendor or fix it yourself, you need to decide who is responsible for correcting the problem so that you will know which vendor to call. In most cases, this means you have to be able to isolate the problem to the specific equipment or circuit causing the trouble.

Common Sense

Common sense is often the most powerful tool you have for isolating a problem. When a problem develops, we suggest you first make a checklist of where the problem could be.

For example, a terminal operator at a remote site reports that the terminal doesn't work. What could be wrong?

1. The terminal has no power—perhaps it isn't plugged in or has a blown fuse.
2. The modem at the terminal isn't connected to the data circuit or has no power.
3. The operator doesn't know how to use the terminal or run the program.

4. The circuit between the terminal and the computer is broken (that is, there is no continuity) or is out of "spec"—not within the specifications needed to function with the rest of the network equipment.
5. The modem at the computer center isn't connected to the data circuit or has no power.
6. The computer is down.

Then go through the checklist to eliminate as many causes as possible.

1. Does the terminal show indications of power (for example, the CRT has a display, the power light is on, and so on)?
2. Does the modem show indications of power and that it is connected to the data circuit (for example, the modem power light is on, the carrier-detect light is on, and so on)?
3. The supervisor can run the program.
4. Various tests show the circuit has continuity and is carrying data (for example, the carrier-detect light on both modems is on, the remote analog and digital loop-back tests check out, and so on).

This common-sense process of listing the various pieces of equipment and circuits that could be causing problems and then going down the list and eliminating as many causes as possible will help you reduce the problem quickly to a few possible causes.

Once you have limited the potential cause of the problem to a reasonable number of areas, there is a wide variety of test equipment and techniques to help you isolate the problem. However, in order to use any of the test equipment, you need to be able to connect your test equipment to the network.

Patching and Switching

Connecting test equipment to the network is best accomplished through either patching or switching. Patching is the process of manually connecting test equipment to the data circuit. This can be accomplished either in parallel, so the circuit is not disturbed, or the test equipment can temporarily replace the equipment on the tested circuit.

The simplest and probably most effective approach to patching is with a plug-and-jack arrangement. Patch panels can be set up so that testing and monitoring equipment can be plugged into the jacks corresponding to the line being checked without disturbing the equipment or circuit. Patch panels can also be used to substitute a known good piece of equipment, such as a modem, for a bad one.

In very small or simple networks, modems can be unplugged from the circuit by their modular plugs, and new modems or test equipment can be plugged into their place for simple tests.

For larger or more complex networks, automated switching systems consisting of switches and relays can be substituted for plugs and jacks. This makes a cleaner and more manageable network control facility.

Network access systems, such as plug-and-jack panels or switches and relays, are often designed with bridges, such as transformers, for analog circuits so that the switching systems will not introduce problems of their own when utilized.

Diagnostic Tests Requiring No Test Equipment

In most cases, basic troubleshooting techniques can be accomplished with little or no special test equipment. If step-by-step procedures are followed, these techniques usually allow the problem to be isolated quickly and easily.

Using a Speaker or Handset The first test is accomplished with a speaker or handset placed across the line. By listening, you can detect a number of impairments such as:

- A high level of white noise
- A dead line
- Cross talk
- Loss of carrier
- Dial pulsing (that is, noise from a rotary-dial telephone interfering with the data circuit).

Using the Modem's Built-In Diagnostics The next test is to use the diagnostics built into the modems you are using. These may include:

- A data quality monitor (DQM)
- Carrier-detect light
- Transmit-data light
- Receive-data light
- Local analog loop-back test
- Local digital loop-back test
- Remote digital loop-back test

The *data quality monitor* consists of test circuitry that analyzes the distortion of the data. If the distortion is greater than a certain preset level, a warning light or alarm will go on.

The *carrier-detect light* (usually a light-emitting diode, or LED) indicates when the carrier from the far-end modem is being received by the modem you are looking at. If the carrier-detect light is on, then you have circuit continuity in that direction, and the far-end modem is transmitting the carrier tone.

The *transmit-data light* flashes in time with the data being transmitted. This tells you when the data are being transmitted and also gives you a relative indication of the data speed. For example, if you have four modems and someone accidentally switched one of them to 2400 bps when they were all supposed to be at 1200 bps, you would see the difference in the transmit-data light flash rates.

The *receive-data light* flashes in time with the data being received. This indicates that you have circuit continuity in the receive direction and that you are receiving data from the remote modem.

If your modem is equipped with loop-back tests, three tests can be performed:

The *local analog loop-back test* takes the analog output from the transmit side of the local modem, loops it back to the analog input or receive side of the modem, and compares what is received with what is transmitted.

The *local digital loop-back test* takes the digital output from the local modem, loops it back through the receive side, and compares it with the digital input.

If both the local digital and analog loop-back tests check out, the odds are that the modem is good.

The *remote digital loop-back test* takes a signal from the local modem, transmits it to the far-end modem where the analog signal is converted to digital form and then looped back to the transmit side of the far-end modem, converted to analog form, and sent back to the local modem where it is received, converted back to digital form, and compared to the original signal sent. This tests both the near- and far-end modems and the transmission facility as well as the electrical interface connectors (such as the RS-232 connectors). However, this test requires a full-duplex modem of either a two-wire or a four-wire type.

Using Test Equipment

If the above tests indicate that the problem is somewhere in the circuit or if they are inconclusive, it may be necessary to use more elaborate test equipment to isolate the problem.

Analog Test Equipment The primary piece of analog test equipment used for troubleshooting analog data is the *line test set*. The line test set can be used to verify the performance of a specific line and to determine if a line is "in spec" with respect to the various parameters that affect data transmission. The line test set can usually perform the following tests:

- Level measurements
- Frequency response measurements
- Noise measurements
- Signal-to-noise ratio measurements
- Equalization measurements.

The exact measurements that you can perform and the specifics of how to perform them will depend on the equipment selected.

Digital Test Equipment A very popular and perhaps one of the most important pieces of test equipment is the *EIA monitor and breakout box (BOB).* It allows you to monitor the status of the signals on various pins of the RS-232 connector. It also allows you to interrupt and cross-patch the signals on the various pins of the RS-232 connector. In this way, various cross-connected options can be tried until you "get it right." Then you can use the breakout box temporarily or permanently in the circuit, or you can wire a special cable to solve your problem.

When you select a breakout box, a few features you might consider are:

- Red LEDs (will light for a space signal) and green LEDs (will light for a mark signal) for each lead being monitored
- A detachable RS-232C cable
- All twenty-five switches individually numbered
- A pulse trap feature, which will light an LED until you reset it if a change from space to mark (or vice versa) is detected.

The following examples show several ways in which a breakout box can be used in troubleshooting data communications problems.

1. You have just connected a printer with an RS-232C serial interface to a high-speed modem that is connected by a private four-wire line to a remote computer. You find that you are constantly getting print buffer overflow and are losing information although the print buffer's flow control works fine when the same printer is connected to the local computer.

You put the breakout box between the printer and the modem and see from the LED's that, when the print buffer is full, the printer "drops" the data terminal ready (DTR) pin (that is, the electrical signal on the DTR pin goes from a high or "on" state to a low or "off" state, signifying that the printer is not ready to receive data). Unfortunately, many private-line modems ignore DTR. Therefore, no flow control is being sent back to the remote computer, and it just keeps on transmitting.

Using the breakout box, you can patch the request to send (RTS, pin 4) to DTR (pin 20) at the printer. Using another breakout box, you can patch the data carrier detect (DCD, pin 8) to clear to send (CTS, pin 5) at the computer end. This uses the presence or absence of carrier for flow control. If this approach solves the flow control problem (as it should), then permanent cables with the proper jumper connections can be made up. Be sure to label the special cables clearly and to record what was done and why in your procedures manual.

2. You are connecting a "dumb" VDT at a remote location via a dial-up modem over the switched voice telephone network to a central computer, but nothing happens, although everything seems to check out fine.

In contrast to the previous example where many private-line modems ignore

DTR, most dial-up modems won't work without DTR. You install your breakout box (BOB) between the VDT and the modem and see that the VDT is not raising DTR. Using your BOB, you open the DTR connection and patch it to an "on" signal, thus forcing DTR "on." The terminal now works and you can either put together a special cable or strap DTR "on" in the modem itself.

3. In this example, we use the BOB to loop the data from the transmit side of a dumb VDT back into the receive side so the VDT can be checked out. Use your BOB to patch RTS (pin 4) to CTS (pin 5), RD (pin 3) to TD (pin 2), and DSR (pin 6) to DTR (pin 20) and DCD (pin 8). Characters typed on the VDT should now appear on the screen.

4. A BOB can very quickly identify whether the RS-232C connector on a device is configured as DCE or DTE. Look at the LEDs while transmitting data from the device in question. If the data goes out pin 2, it is wired as DTE. If the data goes out pin 3, it is wired as DCE.

If you don't have a breakout box, a simple inexpensive *voltmeter* can be used to determine if an RS-232C connector is wired as DTE or DCE. Check the voltage between pins 2 and 7 (signal ground) and pins 3 and 7 with no data being transmitted. If the voltage on 2 (but not on 3) is negative (between -3 and -25 volts), then it is DTE. If the voltage on 3 (but not on 2) is negative (-3 to -25 volts), then it is DCE. If both pins 2 and 3 read a negative voltage relative to pin 7 (-3 to -25 volts), then further testing is required. Set the baud rate to 1200 bps or less (preferably 110 or 300 bps) and transmit a character with the voltmeter connected between pins 2 and 7. Repeat this test with the voltmeter connected between pins 3 and 7. The voltage will fluctuate (that is, the voltmeter needle will vibrate) when it is connected to pin 2 for DTE and when it is connected to pin 3 for DCE.

Bit Error Rate Tester Another popular and important piece of digital test equipment is the BERT or bit error rate tester. This device allows you to transmit a series of data bits, loop the transmission back at the other end (often by transmitting a 2713 Hz tone that activates AT&T's 829 loop-back device), and then compare the received bits with those transmitted.

Character Error Rate Tester The CERT is essentially identical to the BERT except that it transmits complete characters. This allows you to check out the network using the character code set your computer uses. The CERT checks character parity and inserts start and stop bits with each character.

Block Error Rate Tester The next level of digital testers is the BLERT or block error rate tester. When BLERTs are used to test terminals, they are usually referred to as a terminal test set. They are very much like CERTs except that they transmit complete messages, not just characters. Thus, a BLERT or terminal test set can test functions associated with specific messages, such as polling messages.

With most modest data communications networks, the equipment and troubleshooting approaches described here will be sufficient to isolate the problem and determine whom to call to get the problem solved. In some situations, more advanced test equipment, which can time signals and display information on a CRT, is necessary along with a trained technician in order to isolate difficult problems.

ADVANCED NETWORK MANAGEMENT

In large complex networks or in smaller networks where down time would be very damaging to the user, complicated and sophisticated computer-controlled network management systems can be employed. These computer-based systems can continuously monitor data lines, gather and record statistics on the use of the data lines, and identify and record any events that are outside of predetermined parameters.

Not only can these systems provide real-time warning and failure alarms but they also can automatically diagnose certain types of problems and greatly simplify the task of tracing other types of problems. They can also be designed to automatically switch good circuits or equipment into the network to replace faulty circuits and equipment. Furthermore, they can provide printed records of network activity and problems. These network activity records can be the basis for billing users for their time on the network.

With the cost of automated test and network management equipment coming down and the tendency to use higher-speed transmission, automated test and network management systems should be considered even by users with relatively small networks.

FINDING AND FIXING BAD EQUIPMENT

One final piece of advice: substituting a spare modem is often the best way of testing a modem suspected of being bad. A spare modem is also the fastest and usually least expensive way of bringing a network back up when a modem fails. Therefore, we usually recommend that, rather than relying on a service company to maintain your modems, buy a spare and replace the bad modem yourself. Then send the bad modem back to the manufacturer for repair.

The same advice applies to multiplexors, sync-to-async converters, error-correction devices, and so on. The cost of these devices continues to go down, and the cost of a service call continues to go up. Also, a service person can take hours or even days to respond to your call, while if you have a spare, you can often fix the problem in minutes.

POINTS TO REMEMBER

Keep in mind that common sense and simple troubleshooting techniques will often be the easiest and fastest way to isolate a problem. It is important to have written, systematized troubleshooting procedures to follow so that no potential problem, no matter how small, is overlooked. Once the trouble-causing component is found, it should be reported promptly to the organization responsible for repairing it. All of this activity should be entered into a trouble log, and specific follow-up times should be set and followed to be sure the problem is solved in a timely manner.

Finally, you should consider stocking spares for all of your data communications equipment and repair a failed unit by replacing it with the spare. I find this is not only much faster but usually less expensive as well.

THE SEND-OFF

Congratulations! You have now completed a very comprehensive review of the concepts, jargon, technology, and techniques of data communications. You should now be familiar with the vocabulary of data communications and able to read and understand data communications publications, to question intelligently different vendors about the features and benefits of their products, and to make meaningful comparisons of the myriad of alternatives that will be offered to you to solve your communications problems. You now know what types of products and solutions to look for—or at least where to begin to look. Finally, you should have picked up enough techniques and suggestions to help you establish or improve your data communications management procedures.

APPENDIX ONE

CODE CHARTS

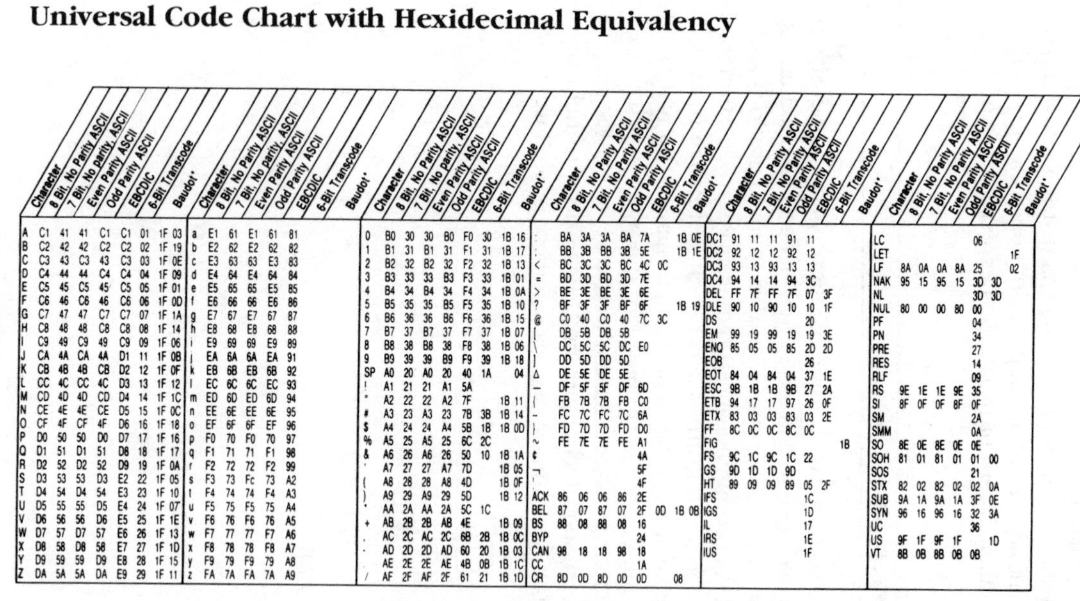

Universal Code Chart with Hexidecimal Equivalency

Character	8 Bit, No Parity ASCII	7 Bit, No parity, ASCII	Even Parity ASCII	Odd Parity ASCII	EBCDIC	6-Bit Transcode	Baudot*
A	C1	41	41	C1	C1	01	1F 03
B	C2	42	42	C2	C2	02	1F 19
C	C3	43	C3	43	C3	03	1F 0E
D	C4	44	44	C4	C4	04	1F 09
E	C5	45	C5	45	C5	05	1F 01
F	C6	46	C6	46	C6	06	1F 0D
G	C7	47	47	C7	C7	07	1F 1A
H	C8	48	48	C8	C8	08	1F 14
I	C9	49	C9	49	C9	09	1F 06
J	CA	4A	CA	4A	D1	11	1F 0B
K	CB	4B	4B	CB	D2	12	1F 0F
L	CC	4C	CC	4C	D3	13	1F 12
M	CD	4D	4D	CD	D4	14	1F 1C
N	CE	4E	4E	CE	D5	15	1F 0C
O	CF	4F	CF	4F	D6	16	1F 18
P	D0	50	50	D0	D7	17	1F 16
Q	D1	51	D1	51	D8	18	1F 17
R	D2	52	D2	52	D9	19	1F 0A
S	D3	53	53	D3	E2	22	1F 05
T	D4	54	D4	54	E3	23	1F 10
U	D5	55	55	D5	E4	24	1F 07
V	D6	56	56	D6	E5	25	1F 1E
W	D7	57	D7	57	E6	26	1F 13
X	D8	58	D8	58	E7	27	1F 1D
Y	D9	59	59	D9	E8	28	1F 15
Z	DA	5A	5A	DA	E9	29	1F 11

Character	8 Bit, No Parity ASCII	7 Bit, No parity, ASCII	Even Parity ASCII	Odd Parity ASCII	EBCDIC	6-Bit Transcode	Baudot*
a	E1	61	E1	61	81		
b	E2	62	E2	62	82		
c	E3	63	63	E3	83		
d	E4	64	E4	64	84		
e	E5	65	65	E5	85		
f	E6	66	66	E6	86		
g	E7	67	E7	67	87		
h	E8	68	E8	68	88		
i	E9	69	69	E9	89		
j	EA	6A	6A	EA	91		
k	EB	6B	EB	6B	92		
l	EC	6C	6C	EC	93		
m	ED	6D	ED	6D	94		
n	EE	6E	EE	6E	95		
o	EF	6F	6F	EF	96		
p	F0	70	F0	70	97		
q	F1	71	71	F1	98		
r	F2	72	72	F2	99		
s	F3	73	Fc	73	A2		
t	F4	74	74	F4	A3		
u	F5	75	F5	75	A4		
v	F6	76	F6	76	A5		
w	F7	77	77	F7	A6		
x	F8	78	78	F8	A7		
y	F9	79	F9	79	A8		
z	FA	7A	FA	7A	A9		

Character	8 Bit, No Parity ASCII	7 Bit, No parity, ASCII	Even Parity ASCII	Odd Parity ASCII	EBCDIC	6-Bit Transcode	Baudot*
0	B0	30	30	B0	F0	30	1B 16
1	B1	31	B1	31	F1	31	1B 17
2	B2	32	B2	32	F2	32	1B 13
3	B3	33	33	B3	F3	33	1B 01
4	B4	34	B4	34	F4	34	1B 0A
5	B5	35	35	B5	F5	35	1B 10
6	B6	36	36	B6	F6	36	1B 15
7	B7	37	B7	37	F7	37	1B 07
8	B8	38	B8	38	F8	38	1B 06
9	B9	39	39	B9	F9	39	1B 18
SP	A0	20	A0	20	40	1A	04
!	A1	21	21	A1	5A		
"	A2	22	22	A2	7F		1B 11
#	A3	23	A3	23	7B	3B	1B 14
$	A4	24	24	A4	5B	1B	1B 0D
%	A5	25	A5	25	6C	2C	
&	A6	26	A6	26	50	10	1B 1A
'	A7	27	27	A7	7D		1B 05
(	A8	28	28	A8	4D		1B 0F
)	A9	29	A9	29	5D		1B 12
*	AA	2A	AA	2A	5C	1C	
+	AB	2B	2B	AB	4E		1B 09
,	AC	2C	AC	2C	6B	2B	1B 0C
-	AD	2D	2D	AD	60	20	1B 03
.	AE	2E	2E	AE	4B	0B	1B 1C
/	AF	2F	AF	2F	61	21	1B 1D

Character	8 Bit, No Parity ASCII	7 Bit, No parity, ASCII	Even Parity ASCII	Odd Parity ASCII	EBCDIC	6-Bit Transcode	Baudot*
:	BA	3A	3A	BA	7A		1B 0E
;	BB	3B	BB	3B	5E		1B 1E
<	BC	3C	3C	BC	4C	0C	
=	BD	3D	BD	3D	7E		
>	BE	3E	BE	3E	6E		
?	BF	3F	3F	BF	6F		1B 19
@	C0	40	C0	40	7C	3C	
[	DB	5B	DB	5B			
\	DC	5C	5C	DC	E0		
]	DD	5D	DD	5D			
Δ	DE	5E	DE	5E			
—	DF	5F	5F	DF	6D		
{	FB	7B	7B	FB	C0		
-	FC	7C	FC	7C	6A		
}	FD	7D	7D	FD	D0		
~	FE	7E	7E	FE	A1		
¢					4A		
¬					5F		
'					4F		
ACK	86	06	06	86	2E		
BEL	87	07	87	07	2F	0D	1B 0B
BS	88	08	88	08	16		
BYP					24		
CAN	98	18	18	98	18		
CC					1A		
CR	8D	0D	8D	0D	0D		08

Character	8 Bit, No Parity ASCII	7 Bit, No parity, ASCII	Even Parity ASCII	Odd Parity ASCII	EBCDIC	6-Bit Transcode	Baudot*
DC1	91	11	11	91	11		
DC2	92	12	12	92	12		
DC3	93	13	93	13	13		
DC4	94	14	14	94	3C		
DEL	FF	7F	FF	7F	07	3F	
DLE	90	10	90	10	10	1F	
DS					20		
EM	99	19	99	19	19	3E	
ENQ	85	05	05	85	2D	2D	
EOB					26		
EOT	84	04	84	04	37	1E	
ESC	9B	1B	1B	9B	27	2A	
ETB	94	17	17	97	26	0F	
ETX	83	03	03	83	03	2E	
FF	8C	0C	0C	8C	0C		
FIG							1B
FS	9C	1C	9C	1C	22		
GS	9D	1D	1D	9D			
HT	89	09	09	89	05	2F	
IFS					1C		
IGS					1D		
IL					17		
IRS					1E		
IUS					1F		

Character	8 Bit, No Parity ASCII	7 Bit, No parity, ASCII	Even Parity ASCII	Odd Parity ASCII	EBCDIC	6-Bit Transcode	Baudot*
LC					06		
LET							1F
LF	8A	0A	0A	8A	25		02
NAK	95	15	95	15	3D	3D	
NL					3D	3D	
NUL	80	00	00	80	00		
PF					04		
PN					34		
PRE					27		
RES					14		
RLF					09		
RS	9E	1E	1E	9E	35		
SI	8F	0F	0F	8F	0F		
SM					2A		
SMM					0A		
SO	8E	0E	8E	0E	0E		
SOH	81	01	81	01	01	00	
SOS					21		
STX	82	02	82	02	02	0A	
SUB	9A	1A	9A	1A	3F	0E	
SYN	96	16	96	16	32	3A	
UC					36		
US	9F	1F	9F	1F		1D	
VT	8B	0B	8B	0B	0B		

*The listing in the Baudot column may be in the form of two hex numbers, in which case the first number designates the LETTERS/FIGURE SHIFT. Where only one hex number is shown, the character is shift independent.

Seven-Bit ASCII Code Chart (Parity Bit Not Shown)

Bits	b4	b3	b2	b1	column ► / row ▼	0	1	2	3	4	5	6	7
b7 →						0	0	0	0	1	1	1	1
b6 →						0	0	1	1	0	0	1	1
b5 →						0	1	0	1	0	1	0	1
	0	0	0	0	0	NUL	DLE	SP	0	@	P		p
	0	0	0	1	1	SOH	DC1	!	1	A	Q	a	q
	0	0	1	0	2	STX	DC2	″	2	B	R	b	r
	0	0	1	1	3	EXT	DC3	#	3	C	S	c	s
	0	1	0	0	4	EOT	DC4	$	4	D	T	d	t
	0	1	0	1	5	ENQ	NAK	%	5	E	U	e	u
	0	1	1	0	6	ACK	SYN	&	6	F	V	f	v
	0	1	1	1	7	BEL	ETB	/	7	G	W	g	w
	1	0	0	0	8	BS	CAN	(	8	H	X	h	x
	1	0	0	1	9	HT	EM	)	9	I	Y	i	y
	1	0	1	0	10	LF	SUB	*	:	J	Z	j	z
	1	0	1	1	11	VT	ESC	+	;	K	[	k	{
	1	1	0	0	12	FF	FX	,	<	L	/	l	|
	1	1	0	1	13	CR	GS	–	=	M	]	m	}
	1	1	1	0	14	SO	RS	.	>	N	>	n	~
	1	1	1	1	15	SI	US	/	?	O	—	°	DEL

Col/Row		Mnemonic and Meaning*	Col/Row		Mnemonic and Meaning*
0/0	NUL	Null	1/0	DLE	Data Link Escape (CC)
0/1	SOH	Start of Heading (CC)	1/1	DC1	Device Control 1
0/2	STX	Start of Text (CC)	1/2	DC2	Device Control 2
0/3	ETX	End of Text (CC)	1/3	DC3	Device Control 3
0/4	EOT	End of Transmission (CC)	1/4	DC4	Device Control 4
0/5	ENQ	Enquiry (CC)	1/5	NAK	Negative Acknowledge (CC)
0/6	ACK	Acknowledge (CC)	1/6	SYN	Synchronous Idle (CC)
0/7	BEL	Bell	1/7	ETB	End of Transmission Block (CC)
0/8	BS	Backspace (FE)	1/8	CAN	Cancel
0/9	HT	Horizontal Tabulation (FE)	1/9	EM	End of Medium
0/10	LF	Line Feed (FE)	1/10	SUB	Substitute
0/11	VT	Vertical Tabulation (FE)	1/11	ESC	Escape
0/12	FF	Form Feed (FE)	1/12	FS	File Separator (IS)
0/13	CR	Carriage Return (FE)	1/13	GS	Group Separator (IS)
0/14	SO	SHift Out	1/14	RS	Record Separator (IS)
0/15	SI	Shift In	1/15	US	Unit Separator (IS)
			7/15	DEL	Delete

*(CC) Communication Control; (FE) Format Effector; (IS) Information Separator

Eight-Bit EBCDIC Code Chart for IBM 3270 Terminals

Bits 4567	Hex1	00				01				10				11			
		00	01	10	11	00	01	10	11	00	01	10	11	00	01	10	11
		0	1	2	3	4	5	6	7	8	9	A	B	C	D	E	F
0000	0	NUL	DLE			SP	&	-									0
0001	1	SOH	SBA					/		a	j			A	J		1
0010	2	STX	EUA		SYN					b	k	s		B	K	S	2
0011	3	ETX	IC							c	l	t		C	L	T	3
0100	4									d	m	u		D	M	U	4
0101	5	PT	NL							e	n	v		E	N	V	5
0110	6			ETB						f	o	q		F	O	W	6
0111	7			ESC	EOT					g	p	x		G	P	X	7
1000	8									h	q	y		H	Q	Y	8
1001	9		EM							i	r	z		I	R	X	9
1010	A					¢	!	|	:								
1011	B					.	$	.	#								
1100	C		DUP		RA	⟨	.	%	@								
1101	D		SF	ENQ	NAK	(	)	_	.								
1110	E		FM			+	;	⟩	=								
1111	F		ITB		SUB	|	—	?	"								

← Bits 0,1
← 2,3
← Hex 0

APPENDIX TWO—INTERFACE PINOUT AND EQUIVALENCY TABLE

Pinout Table for RS-449 and RS-232C/V.24 Interfaces

RS-449 Interface					RS-232 Interface				Signal Type & Direction						
	37 PIN								GND	DATA		CONTROL		TIMING	
9 PIN AUX.	A	B	RS449 CIRCUIT	RS449 DESCRIPTION	25 PIN	EIA-RS232C CIRCUIT	CCITT-V.24 CIRCUIT	RS232 DESCRIPTION		From DCE	To DCE	From DCE	To DCE	From DCE	To DCE
1	1			Shield	1	AA	101	Protective Ground	X						
5	19		SG	Signal Ground	7	AB	102	Signal Ground/Common Return	X						
9	37		SC	Send Common			102a	DTE Common	X						
6	20		RC	Receive Common			102b	DCE Common	X						
	4	22	SD	Send Data	2	BA	103	Transmitted Data			X				
	6	24	RD	Receive Data	3	BB	104	Received Data		X					
	7	25	RS	Request to Send	4	CA	105	Request to Send					X		
	9	27	CS	Clear to Send	5	CB	106	Clear to Send				X			
	11	29	DM	Data Mode	6	CC	107	Data Set Ready				X			
	12	30	TR	Terminal Ready	20	CD	108.2	Data Terminal Ready					X		
	15		IC	Incoming Call	22	CE	125	Ring Indicator				X			
	13	31	RR	Receiver Ready	8	CF	109	Received Line Signal Detector				X			
	33		SQ	Signal Quality	21	CG	110	Signal Quality Detector				X			
	16		SR	Signaling Rate Selector	23	CH	111	Data Signal Rate Selector (DTE)					X		
	2		SI	Signaling Rate Indicator	23	CI	112	Data Signal Rate Selector (DCE)				X			
	17	35	TT	Terminal Timing	24	DA	113	Transmitter Signal Element Timing (DTE)							X
	5	23	ST	Send Timing	15	DB	114	Transmitter Signal Element Timing (DCE)						X	
	8	26	RT	Receive Timing	17	DD	115	Receiver Signal Element Timing (DCE)						X	
3			SSD	Secondary Send Data	14	SBA	118	Secondary Transmitted Data			X				
4			SRD	Secondary Receive Data	16	SBB	119	Secondary Received Data		X					
7			SRS	Secondary Request to Send	19	SCA	120	Secondary Request to Send					X		
8			SCS	Secondary Clear to Send	13	SCB	121	Secondary Clear to Send				X			
2			SRR	Secondary Receiver Ready	12	SCF	122	Secondary Received Line Signal Detector				X			
	10		LL	Local Loopback			141	Local Loopback					X		
	14		RL	Remote Loopback			140	Remote Loopback					X		
	18		TM	Test Mode			142	Test Indicator				X			
	32		SS	Select Standby			116	Select Standby					X		
	36		SB	Standby Indicator			117	Standby Indicator				X			
	16		SF	Select Frequency			126	Select Transmit Frequency				X			
	28		IS	Terminal in Service									X		
	34		NS	New Signal									X		

NOTES:
References DB25 (25 pin) connector that is commonly used for RS232 & V.24. Pins 9 & 10 are reserved for data set testing. Pins 11, 18 and 25 are undefined.
Pins 3 & 21 of RS449 interface connector are undefined.
Lead #23 of RS-232 connector may be defined as CH or CI.
37 pin designation B = return.

RS-449 Interface

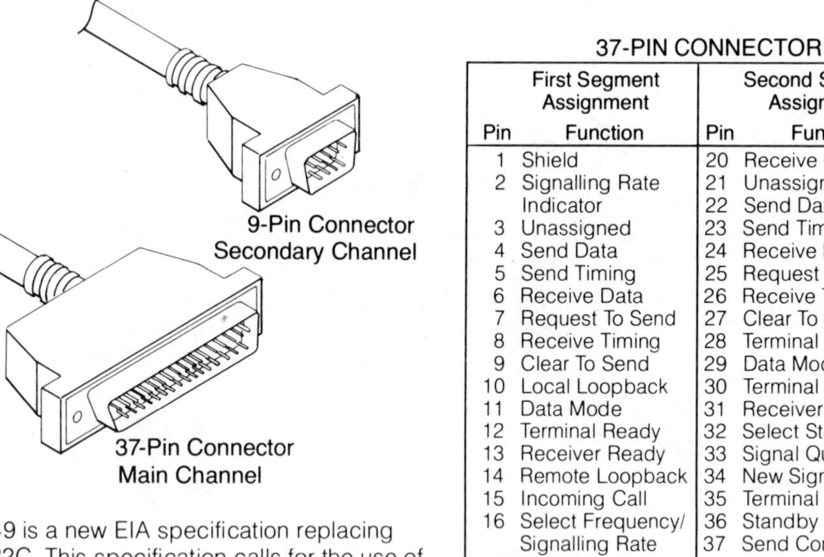

RS449 is a new EIA specification replacing RS232C. This specification calls for the use of a 37-pin connector. For those devices using a side, forward, reverse or secondary channel, a second 9-pin connector is specified. RS449 provides for additional control and signaling.

37-PIN CONNECTOR				9-PIN CONNECTOR	
First Segment Assignment		Second Segment Assignment			
Pin	Function	Pin	Function	Pin	Function
1	Shield	20	Receive Common	1	Shield
2	Signalling Rate Indicator	21	Unassigned	2	Sec. Receiver Ready
		22	Send Data	3	Sec. Send Data
3	Unassigned	23	Send Timing	4	Sec. Receive Data
4	Send Data	24	Receive Data	5	Signal Ground
5	Send Timing	25	Request To Send	6	Receive Common
6	Receive Data	26	Receive Timing	7	Sec. Request To Send
7	Request To Send	27	Clear To Send		
8	Receive Timing	28	Terminal in Service	8	Sec. Clear To Send
9	Clear To Send	29	Data Mode	9	Send Common
10	Local Loopback	30	Terminal Ready		
11	Data Mode	31	Receiver Ready		
12	Terminal Ready	32	Select Standby		
13	Receiver Ready	33	Signal Quality		
14	Remote Loopback	34	New Signal		
15	Incoming Call	35	Terminal Timing		
16	Select Frequency/ Signalling Rate Selector	36	Standby Indicator		
		37	Send Common		
17	Terminal Timing				
18	Test Mode				
19	Signal Ground				

RS-232C/V.24 Interface

Pin	Function	Pin	Function	Pin	Function
1	Frame Ground	10	Negative dc Test Voltage	19	Sec. Request To Send
2	Transmitted Data	11	Unassigned	20	Data Terminal Ready
3	Received Data	12	Sec. Data Carrier Detect	21	Signal Quality Detect
4	Request To Send	13	Sec. Clear To Send	22	Ring Indicator
5	Clear To Send	14	Sec. Transmitted Data	23	Data Rate Select
6	Data Set Ready	15	Transmitter Clock	24	Ext. Transmitter Clock
7	Signal Ground	16	Sec. Received Data	25	Busy
8	Data Carrier Detect	17	Receiver Clock		
9	Positive dc Test Voltage	18	Receiver Dibit Clock		

The terminal connection to the modem is defined by the Electronic Industries Association (EIA) specification RS232C. RS232C specifies the use of a 25-pin connector and the pin on which each signal is placed.

APPENDIX THREE—INTERFACE IDENTIFICATION CHARTS

Interface	Connector Type	Connector
RS-232	DB25	
RS-449	DB37	
449 Secondary, ATARI, DAA, & others	DB9	
TI, NCR POS, & others	DB15	
V.35	V.35	
Data Products, UNIVAC, DEC, & others	M/50	

Interface	Connector Type	Connector
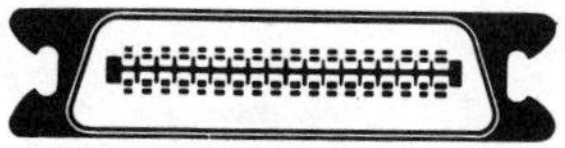 Centronics, Champ, Printronics, Epson & others	Champ	
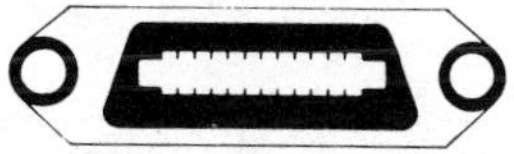 IEEE-488	IEEE-488	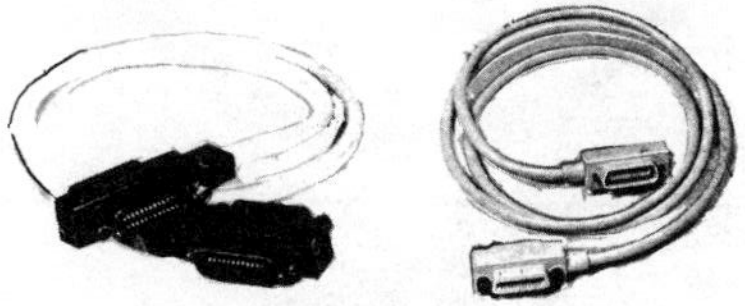
 Current Loop	MATE-N-LOK	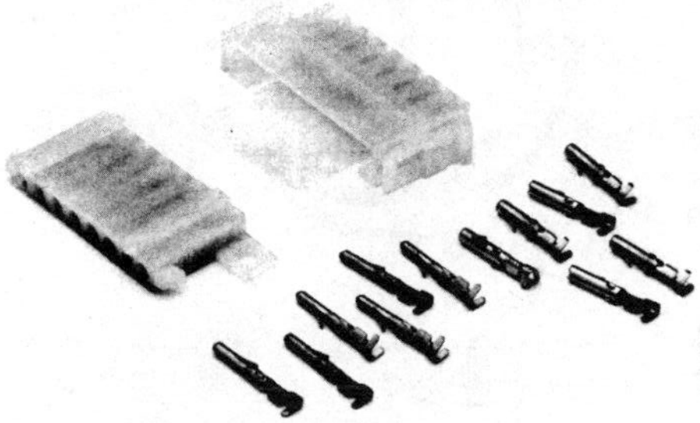
Telephone	RJ-11	

Interface		Connector Type	Connector
Telephone		RJ-45	
BNC	WANG	TNC & BNC	BNC TNC
TNC	Dual Coax.	TNC or BNC	
IBM System 34, 38, 5520 Twinaxial		Twinaxial	

APPENDIX FOUR

INTERFACE VOLTAGE LEVELS

RS-232C Voltage Levels

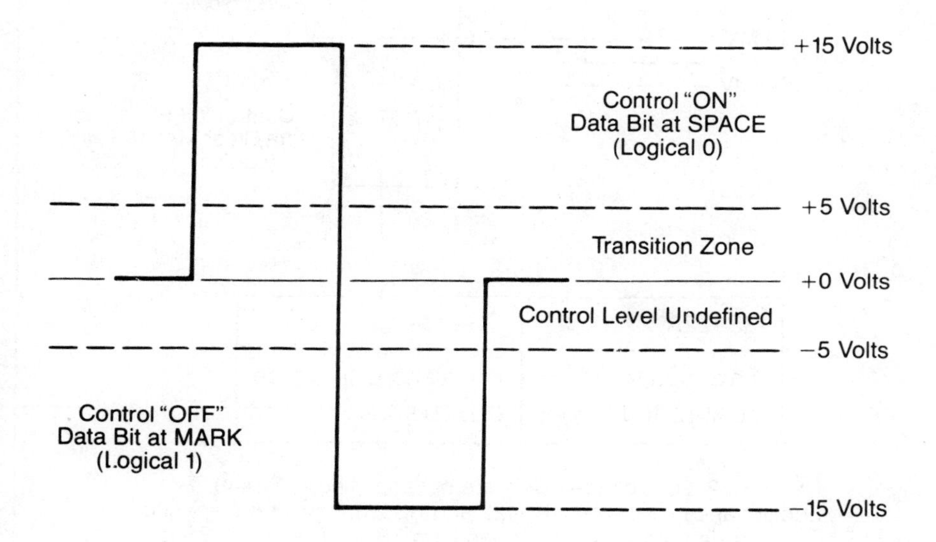

RS232C also defines the level and polarity of the signals going to and from the modem. A logical "1" is referred to as a "MARK" and a logical "0" is called "SPACE." MARK and SPACE are telecommunication terms dating back to Morse code key sets.

RS-449 - RS-442A/RS-423A Voltage Levels

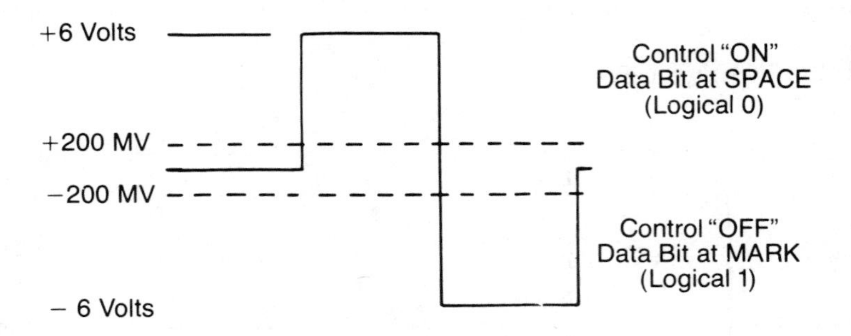

RS422A	RS423A
Balanced Interface 20 kbps to 10 mbps	Unbalanced Interface 0 to 20 kbps

Unlike RS232C, voltage levels are not specified in RS449. Two additional specifications, RS422 and RS423, covering a specific range of data speeds, have been introduced. The use of RS422, RS423, and RS449 also provides for increasing the distance between modems and terminal devices from 50 feet to 4000 feet.

V.28 Voltage Levels (CCITT Recommendation Similar to RS-232C)

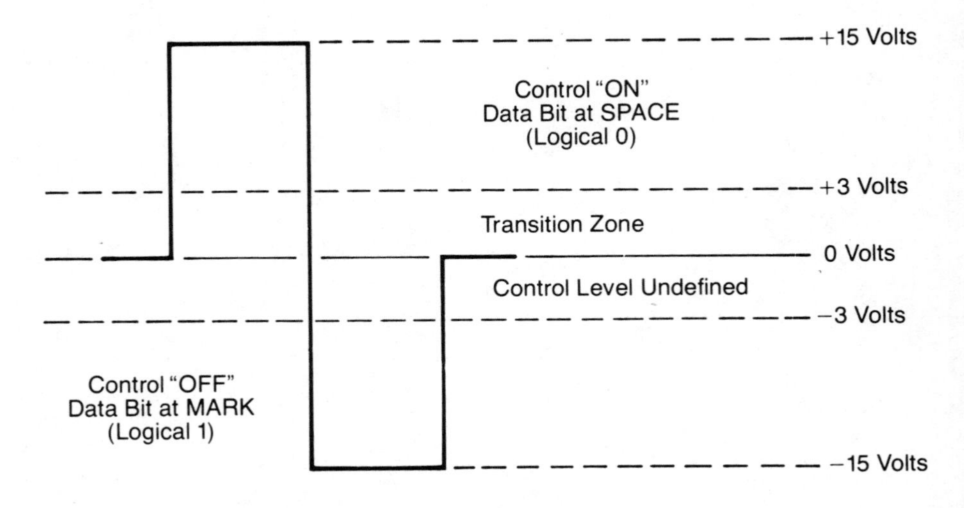

APPENDIX FIVE—AUTOMATIC CALLING UNIT INTERFACES

RS-366 Interface

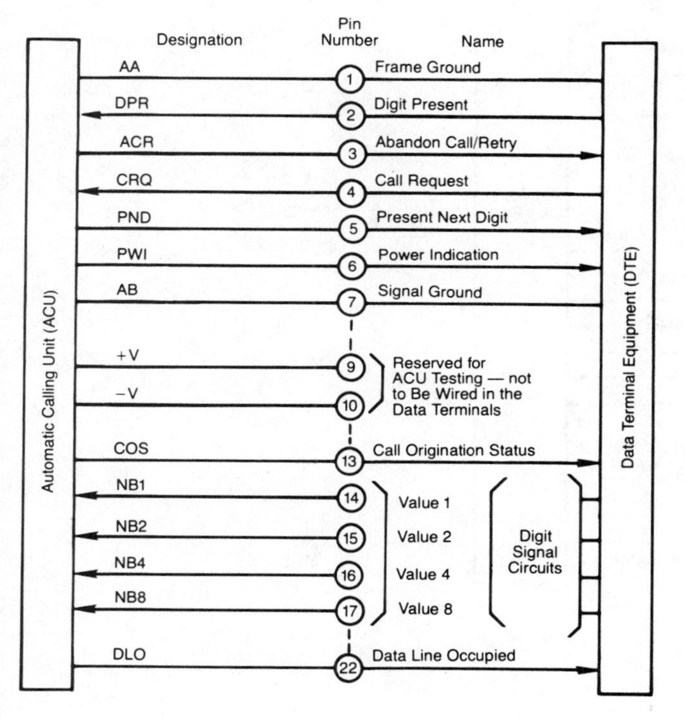

The interface to the computer uses the same type 25-pin connector as RS232 but pin assignments and their functions are different. The RS366 specification outlines the function and pin assignment.

RS-366A Interface

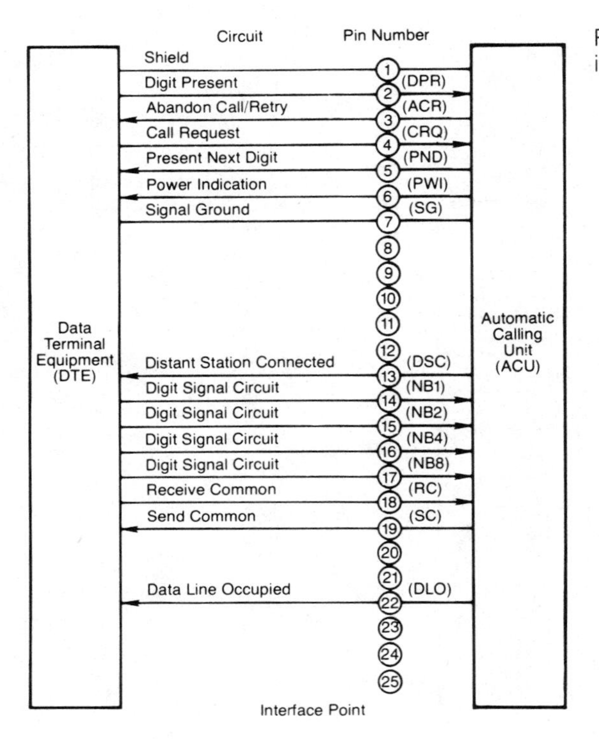

RS366A, a new version of RS366, includes some circuit enhancements.

V.25 Interface

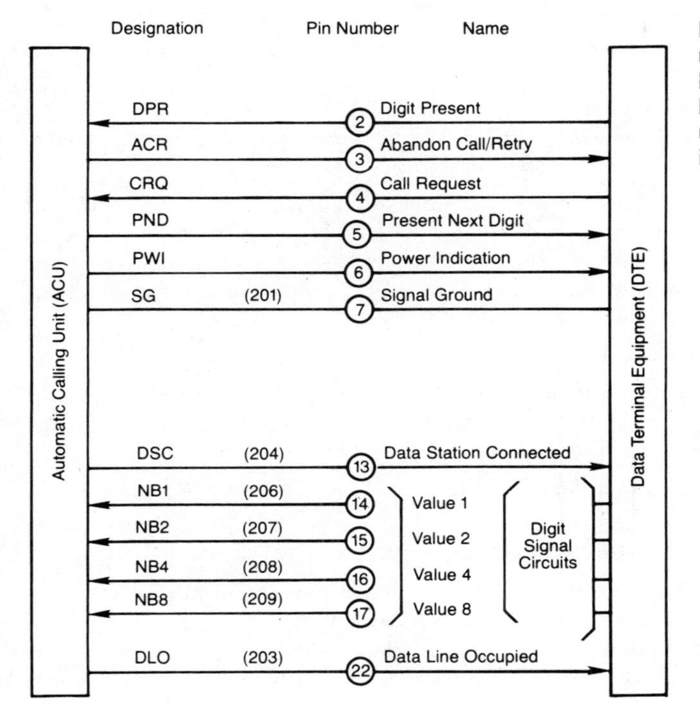

Note 1: Pins not shown are unassigned and should not be used. CCITT Recommendation V.25 is for use with the ACUs and international-type modems.

GLOSSARY

ACK. A control character standing for "acknowledged" or "yes" or "ok", or "no detectable errors" (ch. 4).

Acoustical Coupler. A conventional modem that connects to the telephone line by placing the telephone handset in "cups" on the acoustical coupler, which "couples" or connects the modem with the telephone line acoustically—that is, by allowing the tones or modulated carrier from the modem to pass through the telephone handset's receiver and transmitter just as a human voice would (ch. 5).

ACU. *See* **Automatic Calling Unit.**

Analog Transmission. A technique whereby an information carrier (such as an electronic pulse, a constant voltage, or a radio signal) is modulated so the carrier represents or becomes the analog of the information being transmitted (ch. 2).

ARPANET. An intelligent communications network developed by the U.S. Department of Defense (ch. 7).

ARQ. *See* **Automatic Repeat Request.**

ARQ, Continuous; Go-Back-N-Technique; Selective Retransmission. An error-correction technique whereby the sending device sends data continuously without waiting for an ACK. If an error is detected, the sending device either resends the message in error (Selective Retransmission) or resends the message in error plus all messages that were sent after it (Go-Back-N-Technique) (ch. 4).

ARQ, Stop and Wait. An error-correction technique whereby the sending device transmits one character or message at a time, then stops and waits for an ACK or NAK before transmitting the next data or repeating the last data if that contained an error. (ch. 4). *See also* **ARQ, Continuous; Automatic Repeat Request.**

ASCII. The American National Standard Code for Information Interchange, a 7-bit code that uses an eighth bit per group for parity. The 7 bits allow for 128 unique symbols. ASCII is used extensively with asynchronous transmission, especially with mini and micro computer systems (ch. 3).

Asynchronous Transmission. A character-oriented transmission scheme whereby each character is framed by a start and stop bit, which in turn allows each character to be sent at random intervals or asynchronously (ch. 3).

Attenuation Distortion—Frequency Response. A type of distortion whereby the attenuation of the signal is greater at higher frequencies than at lower frequencies (ch. 5).

Automatic Calling Unit (ACU). A device that under the control of a computer or computer terminal automatically seizes a telephone line and dials the desired number (ch. 3).

Automatic Repeat Request (ARQ). An error-correction scheme in which if an error is detected, the transmitting device is automatically requested to repeat the message that contained the error (ch. 4). *See also* **Forward Error Correction; ARQ, Stop and Wait; ARQ, Continuous.**

Bandwidth. The difference between the highest and lowest frequencies that can be transmitted over a band or transmission facility. Thus, if a facility can transmit between 300 and 3300Hz it has a bandwidth of 3000Hz (ch. 2). *See also* **Frequency Range.**

Baseband. A transmission technique in which the transmission medium carries one and only one signal at a time (ch. 10). *See also* **Broadband.**

Baud. A measure of the speed of a data communications line. The term refers to the number of signal elements, or signal events, passing a point on the line per second. If each signal element represents 3 bits and 1600 signal elements pass a point on the line each second, then the data is being transmitted at 1600 Baud but at 4,800 bits per second (ch. 3).

Baudot Code. A code that uses five bits per group to directly represent thirty-two different symbols. However, each group of five bits is interpreted to be one of two possible symbols depending on whether the group of bits follows a "letters" or "figures" control character. This then allows for sixty-two symbols, and is called a sequential code (ch. 3).

Baud Rate. The peak speed at which data is being transmitted, expressed in Baud (ch. 3).

BCC (Block Check Characters). A control character used to check whether a block of data contains any errors (ch. 4). *See also* **Longitudinal Redundancy Check.**

BERT, Bit Error Rate Tester. A piece of digital test equipment that transmits a series of data bits, loops the transmission back at the other end, and then compares the received bits with those transmitted. A "CERT" or "Character Error Rate Tester" is essentially the same as a "BERT" but transmits complete characters instead of bits. A "BLERT" or "Block Error Rate Tester" is essentially the same as a "CERT" but sends complete messages or blocks instead of just characters (ch. 13).

Binary Synchronous (BISYNC). A family of character-oriented, non-positioned protocols utilizing synchronous transmission, developed by IBM primarily for communications between a host computer and its terminals (ch. 4).

Bit. A contraction of "BInary digiT", either a 0 or a 1 (ch. 3).

Bit Serial Interface. A hardware interface in which a series of bits is transmitted one after another over a single channel (ch. 3). *See also* **Parallel Interface.**

Bit, Start, and Stop. In asynchronous transmission this signals bits added to the beginning and end of each character to provide character synchronization (ch. 3).

Bit Striping. Removing the start-stop bits on each ASYNC character and transmitting the data using some type of synchronous technique. This is commonly done with statistical multiplexors which then add back the start-stop bits after demultiplexing the data (ch. 8).

Black Box. A device with an input and output (or multiple inputs and/or outputs) that performs certain functions on the input(s) before outputing the result(s). It is called a black box because we are interested in *what* it does, not in seeing inside the box to learn *how* it does it (ch. 5).

BLERT. Block Error Rate Tester (ch. 13). *See* **BERT.**

Bps. Bits per second. The peak speed at which data is being transmitted, expressed as the number of bits passing a point on the line each second (ch. 3). *See also* **Baud.**

Breakout Box (BOB). Also known as an "EIA monitor", a piece of digital test equipment that monitors the status of the signals on the various pins of an RS232C connector and allows signals to be broken, patched, or cross-connected (ch. 13).

Bridging. Connecting two circuits together (ch. 7).

Broadband. A transmission technique in which a number of different signals can be transmitted over the transmission medium at one time by a technique such as frequency division multiplexing, which breaks the medium's bandwidth into smaller bands for each transmission channel (ch. 10). *See also* **Baseband.**

Buffering. Storing data temporarily (ch. 3).

Bus. A data communications highway to which numerous devices can be connected, usually along its entire length (ch. 9).

Bus Network. A long line or bus with the computer(s) and peripherals attached or "dropped" at various points. The bus is therefore a form of multidrop or multipoint network (ch. 5).

Byte. A group of bits processed as a unit, usually a group of 8 bits that can represent up to 256 different symbols (ch. 3).

Carterfone Decision. An FCC decision issued June 26, 1968, authorizing customers to connect their own equipment to the telephone company's lines (ch. 11).

CCITT (Consultative Committee for International Telegraphy and Telephony). A subcommittee of the International Telecommunications Union (ITU), which, in turn, is an agency of the United Nations (ch. 3).

C-Conditioning. A family of services offered by the telephone company that controls attenuation distortion and envelope delay distortion on lines used for data communications (ch. 5).

Cellular Radio. A technique to provide mobile radio service by passing along the conversation under computer control to different transmitters as the mobile radio moves from one cell to another (ch. 9).

Centronics Interface. A thirty-six-pin parallel interface designed by the Centronics Computer Data Corporation that has become a de facto standard for connecting printers to mini and micro computers (ch. 3).

CERT. Character Error Rate Tester (ch. 13). *See* **BERT.**

Channel. A term that is generally used to indicate a one-way means or path for communications and that does not refer to the physical media itself. A full-duplex circuit consists of two channels, one for transmission and one for reception (ch. 3).

Channel, Forward. *See* **Primary Channel.**

Channel, Main. *See* **Primary Channel.**

Channel, Primary or Forward or Main. The higher speed of two channels used for transmitting the data (ch. 3). *See also* **Reverse Channel.**

Channel, Reverse. The slower speed of two channels used for low-volume information such as error-detection acknowledgement (ch. 3). *See also* **Primary Channel.**

Channel Service Unit (CSU). A device required by AT&T in order to connect a user's DTE to a digital transmission line for DDS, if the customer's DTE can transmit data in the specified bipolar format and thus does not require a DSU (ch. 5).

Circuit. Either the physical communications media over which data is transmitted, or one of many two-way communications paths derived from a single physical communications media (ch. 3).

Circuit, Multipoint. A circuit with DTE connected at intermediate points in addition to the two end points (ch. 3).

Circuit, Point-to-Point. A circuit with only two connections (DTE is attached at only two points on the circuit), one at each end (ch. 3).

Circuit, Segment. A single point-to-point circuit, a term usually employed where the circuit is one part of a multipoint circuit or a network (ch. 3). *See also* **Link.**

Circuit, Switched. A circuit that can be connected to any of many different points, usually by dialing a number from a telephone or transmitting a number from a computer or computer terminal (ch. 3).

Clipping. *See* **Non-Linear Distortion.**

Clock. An electronic device that generates the timing signal used to maintain synchronization (ch. 3).

Clock, External. A clock or timing signal provided from another device; a modem can provide an external clock to a terminal (ch. 3).

Cluster Controller. A device that consolidates some of the data communication functions that would otherwise have to be duplicated in each of the devices connected to it (ch. 6).

COAX. *See* **Coaxial Cable.**

Coaxial Cable (COAX). Consists of one wire in the center surrounded by insulation and covered by a wire mesh or foil shield. Coaxial cables can carry a very broad band signal and thus carry much more data than twisted pairs (ch. 9).

Code. Groups of bits and their uniquely associated symbols. Most codes assign groups of bits to the digits 0 through 9, the letters of the alphabet, and specialized characters such as control characters (ch. 3).

Codec. A contraction of "COder" and "DECoder", electronic circuits that code and decode analog voice signals into and from digital signals (ch. 2).

Common Carriers. With regard to communications, companies (such as telephone or telegraph companies) that provide communications services to the general public (ch. 11).

Communications Line Controller. The piece of computer hardware which, along with its associated software, accepts data from the keyboard or memory and performs the necessary functions to pass that data on to the modem—such as converting each character to a stream of bits (ch. 3).

Concentrator. A device that combines data from several other devices onto one circuit similar to a multiplexor but usually with more features such as data compression, forward-error correction, selective routing, and buffering of long messages (ch. 8).

Constant Ratio Code or X-Out-Of-Y Code. An infrequently used mathematical approach to error detection (ch. 4).

Contention. A network access method whereby all devices on the line try to access that line whenever they have a message to send. The most popular technique used to avoid interference in contention systems is CSMA (ch. 6).

Control Characters. Special characters or a group of bits that have a special meaning for the process being controlled (ch. 3).

CRC. *See* **Cyclic Redundancy Check.**

CRT. *See* **Video Display Terminal.**

CSMA, Carrier Sense Multiple Access. A method used to police traffic in contention systems. When a device wants to send data, it listens for a tone or carrier. If it hears no carrier, it transmits a carrier tone and then transmits the data. Since this technique can lead to overlapping data or collisions, more sophisticated CSMA systems utilize a technique to detect collisions and then retransmit data. These are called "CSMA/CD" or "Carrier Sense Multiple Access with Collision Detection" systems (ch. 6).

CSMA/CD. *See* **CSMA.**

CSU. *See* **Channel Service Unit.**

Cyclic Redundancy Check (CRC). An error-detection scheme utilizing a complicated mathematical formula to generate a very accurate check on the data being received (ch. 4).

DAA. *See* **Data Access Arrangement.**

Data Access Arrangement (DAA). A device required by telephone company tariffs to connect a customer-supplied DCE that is not FCC-registered to the telephone network. Any modem registered with the FCC as meeting FCC requirements may be connected without the use of a DAA (ch. 5).

Data Bank. A centralized collection of various types of computerized information accessible by means of dial-type access lines to various users at diverse locations (ch. 1).

Data Circuit Terminating Equipment. *See* **Data Communications Equipment.**

Data Communications. The process of transmitting information that originates with or is generated by a machine (typically a computer or computer terminal) and that is intended to be received by another machine (ch. 1).

Data Communications Equipment (DCE)., Data Circuit-Terminating Equipment (DCE or DCTE). A term used in data communications to describe collectively the communications equipment that connects the data communications line to the data terminal equipment. DCE usually refers to modems or data sets and sometimes multiplexors (ch. 3).

Data Compression. A term applied to a variety of techniques used to reduce the number of bits necessary to transmit a given set of data (ch. 8).

Data Link. The physical circuit plus all of the equipment necessary for two devices connected to the circuit to communicate. A data link is usually not considered to be established until all handshaking has been completed and all equipment is operating. Do not confuse with "Link" (ch. 3). *See* **Link.**

Data Link Control (DLC). The combination of software and hardware that manages the transmission and receipt of data over the communications line and looks for errors during the process (ch. 4).

Data PBX. A PBX designed to switch data rather than voice. Some modern digital PBXs can function very effectively as both a data PBX and a voice PBX (ch. 8).

Dataphone Digital Service (DDS). One of AT&T's private-line digital transmission services (ch. 9).

Data Processing Terminal Equipment. *See* **Data Terminal Equipment.**

Data Sets. *See* **Modems.**

Data Terminal Equipment (DTE), Data Processing Terminal Equipment (DPTE). A term used in data communications to describe collectively the various data-processing equipment and other sources of data such as computers and communications controllers, and all peripheral equipment such as terminals and printers (ch. 3). *See* **Data Communications Equipment.**

Datagram. If, in a packet-switching network, a message can fit in its entirety into one packet, it is referred to as a datagram and the network does not try to relate that packet to any other packet (ch. 9).

D-Conditioning. A service offered by telephone companies to control the signal-to-noise ratio and harmonic or non-linear distortion. It is primarily used with 9600 bps modems on voice grade private lines (ch. 5).

DC Continuity. A circuit that appears to be a continuous circuit (rather than an open circuit) to a direct current. This continuity is usually accomplished by means of a continuous metallic circuit, that is, a circuit composed of wire that is not interrupted by devices such as amplifiers, transformers, or capacitors (ch. 5).

DCE. *See* **Data Communications Equipment.**

DCTE. *See* **Data Circuit Terminating Equipment.**

DDS. *See* **Dataphone Digital Service.**

DECNET. An intelligent communication network approach developed by Digital Equipment Corporation (ch. 7).

Digital Error. This occurs in digital transmission where a zero is interpreted as a one or vice versa (ch. 2).

Digital Service Unit. A device AT&T refers to as a Data Service Unit (DSU), which connects DTE to a digital transmission line such as AT&T's DDS (ch. 5). *See* **Channel Service Unit.**

Digital Transmission. A technique whereby information (whether data, voice, or video) is coded as a series of 0's and 1's, and those 0's and 1's are transmitted (ch. 2).

Direct Current Loop. A data transmission and interface technique, often used with teleprinters and teletypewriters, which transmits a digital signal usually at 20 milliamperes or 62.5 milliamperes DC (ch. 5).

Distributed Processing. A system whereby several computers are connected, usually over a wide (or distributed) area, so that each computer has access to the data base and peripherals of the other computers in the system (ch. 1).

DLC. *See* **Data Link Control.**

DPTE. *See* **Data Processing Terminal Equipment.**

Drop Out. A type of distortion that results in a decrease in the level of the received signal that is greater than 12db and longer than 4ms (ch. 5) *See* **Gain Hit.**

DSU. *See* **Digital Service Unit.**

DTE. *See* **Data Terminal Equipment.**

Dumb Terminal. From a data communications standpoint this is a terminal that cannot be polled, perform error detection, operate synchronously, use a protocol other than start-stop Async, transmit a formated message, reformat data, add BCCs or sequence numbers, operate in half-duplex with line turnaround, or perform other intelligent data communications functions (ch. 8).

EBCDIC. Extended Binary Coded Decimal Interchange Code is an 8-bit code that allows for 256 unique symbols. EBCDIC is used extensively in IBM mainframe computer applications and does not use a parity bit with each character (ch. 3).

Echo-Back. *See* **Echo Check.**

Echo Check. An error-detecting technique whereby each character having been transmitted to the computer, the computer subsequently transmits the character back to the terminal where it is displayed (ch. 4). Also called **Echo-Plex** and **Echo-Back.**

Echo-Plex. *See* **Echo Check.**

Echo-Suppressor. A device used by telephone companies to control echoes in voice telephone calls (ch. 5).

EIA (Electronics Industry Association). This sets standards for the use of electronics in the U.S. (ch. 3).

Electronic Mail. A system whereby a letter is transmitted electronically to its destination where it can either be displayed electronically on a VDT or printed for final delivery (ch. 1). *See also* **Electronic Message Systems; Electronic Mail Box; Store and Forward Systems.**

Electronic Mail Box. A computer-based system whereby messages or "mail" can be addressed to a recipient. A user checks the mail box by computer entry to see if mail exists for his/her address and, if so, the mail can then be viewed, printed, or electronically manipulated (ch. 1). *See* **Electronic Mail, Electronic Merge System, Store and Forward Systems.**

Electronic Message System. A system whereby a message is transmitted electronically to its destination where it can be displayed, printed, or electronically manipulated (ch. 1). *See* **Electronic Mail, Electronic Mail Box, Store and Forward Systems.**

Encoding Error. Quantization Error. In digital transmission, where the variation in a waveform is lost due to the quantization (coding as a discrete number) inherent in the digital encoding (ch. 2).

ENQ. A control character standing for Enquiry or "Can you accept data?" (ch. 4).

Envelope Delay Distortion. A type of distortion in which the high-frequency components of a signal are delayed by a different amount of time than are low-frequency components. A second aspect of envelope delay distortion is that at higher data speeds certain components of successively transmitted characters can overlap, causing what is called "intersymbol interference" (ch. 5).

Equalizer. A device or circuit used to compensate for envelope delay distortion (ch. 5).

Equalizer, Adaptive. An equalizer that can constantly change its equalization parameters to compensate for changes on the line. This is in contrast to automatic equalization that adjusts only between the RTS and CTS signals (ch. 5). *See* **Training Period.**

Equalizer, Automatic. An equalizer circuit that can adjust itself to compensate automatically for the amount of distortion on the line (ch. 5). *See* **Training Period.**

Equalizer, Fixed. An equalizer circuit that is permanently set at one compromise or average setting (ch. 5).

Equalizer, Manual. An equalizer circuit that must be adjusted or set by the user (ch. 5).

ETB. A control character meaning "End of Text, but more messages to follow." (ch. 4). *See* **ETX.**

Ethernet. An LAN standard established by XEROX, Intel, and DEC, available on a license basis to other companies, which operates over coaxial cables, with a data rate of 10m bps (ch. 10).

ETX. A control character standing for End of Text (ch. 4). *See* **ETB.**

Extended Binary Coded Decimal (EBCD). A code using six bits plus one parity bit per group, originally developed for the IBM selectric typewriters (ch. 3).

Extended Distance Cables. Data communications cables designed to operate at longer distances than standard cables (ch. 9).

Facsimile. *See* **FAX.**

Fast Circuit-Switching Networks. *See* **Fast-connect Networks.**

Fast-connect Networks. These are networks that are designed especially for data communications and that can switch circuits in a fraction of a second (ch. 7). Also known as "Fast Circuit Switching Networks."

FAX, Facsimile. The transmission of pictures electronically (ch. 1).

FCC. *See* **Federal Communications Commission.**

FDM. *See* **Multiplexor, Frequency Division.**

FEC. *See* **Forward Error Correction.**

Federal Communications Commission (FCC). An agency of the U.S. Government that regulates radio and television broadcasters, telephone and data common carriers, as well as other systems in the communications field (ch. 11).

Fixed Loss Loop (FLL) Device. A classification of registered modems that limits their output to −4dBm (ch. 5). *See* **Permissive Device and Programmable Device.**

FLL. *See* **Fixed Loss Loop.**

Flow Control. Protocols or techniques for turning the data transmitter on and off when necessary to control the flow of data, such as when a buffer is full or a printer has run out of paper (ch. 4). *See* **XON—XOFF.**

Forward Error Correction (FEC). An error-correction scheme whereby each message contains sufficient extra information bits, so that when the receiving device detects an error it can use the extra bits to correct that error (ch. 4). *See* **Automatic Repeat Request.**

Four-Wire Circuit. A circuit with separate pairs of wires for transmission and reception (ch. 7).

Frequency Range. The lowest to the highest frequencies that can be transmitted over a band or a facility (ch. 2). *See* **Bandwidth.**

Front-End Communications Processor. A specialized computer dedicated to handling communications functions, thus freeing the host computer to which it is attached for data processing (ch. 3).

Full-Duplex (FDX). A communications term that has several sometimes conflicting meanings: a) When applied to physical circuits "FDX" means a four-wire circuit; b) When applied to a communications channel, protocol, modem, or a transmitting or receiving device "FDX" means it will support communications in both directions simultaneously; c) When used to lable a switch on a terminal HDX/FDX, "FDX" may mean a remote echo mode (ch. 4). *See* **Half-Duplex** and **Simplex.**

Gain Hit. A type of distortion resulting in a sudden increase or decrease in the level of the received signal that is less than 12dB (ch. 5). *See* **Drop Out.**

Gateway. The connecting link (along with its associated communications control software and hardware) between two networks (ch. 6).

Go-Back-N-Technique. *See* **ARQ, Continuous.**

GPIB (General Purpose Interface Bus). This deviates very slightly from the IEEE 488 standard; these are often used interchangeably (ch. 3). *See* **HPIB.**

Half-Duplex (HDX). A communications term that has several sometimes conflicting meanings: a) When applied to physical circuits "HDX" means a two-wire circuit; b) When applied to a communications channel, modem, or a transmitting or receiving device "HDX" means it will support communications in either direction but only one way at a time; however, the International Telecommunications Union defines "Half-Duplex" as a channel that is capable of supporting communications in both directions simultaneously, but, because of the DTE or DCE attached to it, can only operate in one direction at a time; c) When applied to protocols "HDX" means a protocol that can support communications in only one direction at a time; d) When used to lable a switch on a terminal HDX/FDX, "HDX" may mean a local Echo Back mode (ch. 4). *See* **Full-Duplex and Simplex.**

Hardware Interface. The physical hardware that is used to connect electrically the communications line controller to the modem or other DCE (ch. 3).

Harmonic Distortion. *See* **Non-Linear Distortion.**

Hertz (Hz). Another term for cycles per second, as in "300Hz" (ch. 2).

Hierarchical Network Structure. A network structure in which the network's functions are broken down into layers, each layer having a specific and highly defined role or responsibility (ch. 7). *See* **OSI.**

High-Level Data Link Control (HDLC). The standard for advanced data link control protocols adopted by the International Standards Organization (ch. 4). *See* **Synchronous Data Link Control.**

HPIB. Hewlett-Packard Interface Bus. This deviates very slightly from the IEEE 488 standard; they are often used interchangeably (ch. 3). *See* **GPIB.**

IEEE. Institute of Electrical and Electronic Engineers, setting standards for use in the U.S. (ch. 3).

IEEE 488. An IEEE standard for a parallel interface most popularly used to connect test instruments to computers. Sometimes referred to as the instrumentation interface or the IEEE 488 Bus (ch. 3). *See* **GPIB and HPIB.**

Integrated Services Digital Network (ISDN). A new generation of digital communication systems allowing a user to transmit voice, data, video, or any other type of communication that is digitally encoded (ch. 2).

Intelligent Communications Network. A network employing advanced network control hardware and software, which permits alternate routing and other complex network control functions (ch. 7).

Interconnect Industry. The industry that developed following the Carterfone Decision to manufacture, market, and service equipment that connects to the telephone company's lines, as well as related equipment and services (ch. 11).

International Record Carrier (IRC). Common carriers that provide international communications services. They used to provide non-voice services only such as telex, hence the term record carrier (ch. 11).

Intersymbol Interference. *See* **Envelope Delay Distortion.**

IRC. *See* **International Record Carrier.**

ITU. *See* **CCITT.**

LADT (Local Area Data Transport). A telephone company service that moves both voice and 4800 bps data simultaneously over the same telephone line (ch. 9).

Line. This is usually used to refer to a wire circuit, or a circuit obtained from a common carrier (ch. 3).

Line, Dedicated or Private. A line for the exclusive and usually unlimited use of the customer that may be owned by the customer or leased from a common carrier (ch. 3). *See* **Leased Line.**

Line, Dial Up. A line arranged as a switched circuit (ch. 3).

Line Drivers. A term often applied to a variety of devices, sometimes called "short-haul modems" or "limited-distance modems." Line drivers are DCE that transmit or "drive" a digital signal directly across a transmission network, possibly amplifying, reshaping, or coding the digital signal from the DTE before transmitting it (ch. 5).

Line, Leased. A dedicated line leased from a common carrier (ch. 3).

Line Test Set. A piece of analog test equipment that usually measures the following characteristics of a circuit: level, frequency response, noise, signal-to-noise ratio, and equalization (ch. 13).

Link. A single point-to-point circuit, usually used where it is one part of a network (ch. 3). *See* **Circuit Segment and Data Link.**

Local Area Networks (LANs). A medium to high-speed data communications system designed for intrabuilding or intracomple data communications (ch. 10).

Longitudinal Redundancy Check (LRC); and **Block Check Character (BCC).** An error-detection scheme also known as a two-dimensional parity check. In addition to checking for parity in each group of bits, the accuracy of the entire message or block is checked by adding additional bits to the end of the message. These additional bits are called a "BCC" or "Block Check Character" (ch. 4).

Loop. A multipoint network topology consisting of a series of nodes joined by links to form a circle or ring. However, as opposed to a bus network, the communications line in a loop terminates at each node and starts again on the output side of the node. If all nodes in the loop network are equal, that is if there is no controlling node, the network is called a "ring." If there is a controlling node the network is called a "loop" (ch. 6).

Loop-back Tests. A test procedure whereby the output (or transmit) side of a device or communications line is "looped back" or connected to the input (or receive) side, often at a remote site, and a signal is sent out and back again and then compared with the original (ch. 13).

LRC. *See* **Longitudinal Redundancy Check.**

Mark. A term used in asynchronous data transmission, originating from its use in telegraphy: A stop bit, a binary 1, or the rest time between two characters (ch. 3). *See* **Space.**

Mesh Network. A network with links connecting each device to other devices rather than each device only to one central controller. (ch. 6).

Metallic Circuit. *See* **DC Continuity.**

Microprocessor. A computer's central processing unit (CPU) on a semiconductor integrated circuit chip (ch. 1).

Modem. A contraction of "MODulator—DEModulator," data communications equipment that accepts data in the form of digital signals, converts those signals into analog ones that can be transmitted over voice telephone lines, and reconverts the analog signals into digital ones on the receiving end. When a modem is supplied by a telephone company it is often referred to as a Data Set (ch. 3, 5). *See* **Line Drivers.**

Modem, Conventional; Modem, Long Haul. These are true modems which actually modulate and demodulate a signal. They are the ones normally used on voice grade dial-up and leased circuits (ch. 5).

Modem Eliminator, also called NULL modem. A device designed to be inserted between and connecting two data terminal devices. It performs the necessary lead or signal conversions and may be able to simulate the control signals from a modem (ch. 5).

Modem, Fast Turnaround. *See* **Modem, Quick Turnaround.**

Modem, High Speed. A modem operating on voice-grade circuits faster than 4800 bps. This generally means 7200 or 9600 bps modems, although some high-speed modems operate at 14,400 or 16,000 bps (ch. 5).

Modem, Limited Distance. A term usually applied to a variation of the conventional long-haul modem. The limited distance modem lacks some of the latter's electronic functions such as auto-equalization (ch. 5). *See* **Line Drivers.**

Modem, Long Haul. *See* **Modem, Conventional.**

Modem, Low Speed. A modem operating at 1800 bps or less, usually asynchronous (ch. 5).

Modem, Medium Speed. A modem operating at 2000 bps to 4800 bps, usually synchronous (ch. 5).

Modem, Multiport. Also known as a "split stream modem", this is a device that combines a multiplexor and a modem enabling two or more DTEs to be connected to the same line (ch. 5).

Modem, Null. *See* **Modem, Eliminator.**

Modem, Quick Turnaround. Also known as a "fast turnaround modem," this modem continuously sends signals that keep the echo suppressors disabled, thus eliminating a lengthy turnaround time when the line is used in a half-duplex mode (ch. 5).

Modem, Short Haul. A term inconsistently applied to both line drivers and limited-distance modems (ch. 5).

Modem, Split Stream. *See* **Modems, Multiport.**

Modem, Wide Band. A modem designed to operate at speeds greater than those used with high speed modems, such as 19.2k bps, 56k bps, and 230.4k bps. Wide band modems will not operate over voice grade circuits but require a special and more expensive wide band circuit (ch. 5).

MTBF, Mean Time Between Failure. The average expected time between successive failures for a piece of equipment (ch. 12).

Multidrop Network. *See* **Multipoint Network.**

Multiplexor (MUX). A device that combines or mixes a number of data channels onto one circuit by electronic means (ch. 8).

Multiplexor, Frequency Division (FDM). A multiplexor that divides the transmission bandwidth into smaller bandwidths and allocates those to each data channel (ch. 8).

Multiplexor, Inverse. A device that takes a data stream and splits it so the data can be carried over two or more lower capacity channels (ch. 8).

Multiplexor, Statistical (STAT MUX). An intelligent TDM that reassigns unused time slots assigned to a data channel so they can be used by other channels (ch. 8).

Multiplexor, Time Division (TDM). A multiplexor that creates a digital data system and breaks the stream into units or slices and allocates those slices to the various data channels (ch. 8).

Multipoint Network. Also known as a "Multidrop Network", this is a network whereby more than two devices are attached to the line. The bus, tree, and loop topologies are examples of multipoint networks (ch. 6).

MUX. *See* **Multiplexor.**

NAK. A control character standing for "Negative Acknowledgement," or "No", or "Not OK", or "error detected" (ch. 4).

Net Data Throughput (NDT). The number of usable data characters or bits received per second. This does not include control characters or characters received with errors that must be retransmitted (ch. 3).

Network. A connected system of circuits and their associated data communications control equipment (ch. 3).

Network. The transmission system and the associated control software and hardware used to connect computers and peripherals or other computers to each other (ch. 6).

Non-Linear Distortions. Also called "clipping" or "third harmonic distortion", a type of distortion resulting from the fact that the attenuation of a signal varies with its height (ch. 5).

Office Automation. The use of computers to automate hitherto manual office tasks such as typing, filing, sorting, retrieving information, budgeting, forecasting, drafting, and copying (ch. 10).

Optical Fibers. Very thin glass rods that can be used to carry data encoded on light signals. Optical fibers have an even wider bandwidth than coaxial cable and are virtually immune from noise caused by electro-magnetic radiation; they are also virtually immune from unauthorized tapping (ch. 9).

OSI, Open Systems Interconnection Reference Model. A seven-layer hierarchical network structure developed by the International Standards Organization (ch. 7). *See* **Hierarchical Network Structure.**

Other Common Carrier (OCC). Common carriers other than AT&T and the telephone operating companies that are usually referred to as "the common carriers." Prior to the time other common carriers were allowed to offer dial-up telephone service to the general public (they were restricted to private-line service or value-added service), they were known as **specialized common carriers (SCC)** (ch. 11).

Packet-Switching Networks. A class of value-added networks that break the data down into short messages called packets before transmitting the data (ch. 9).

PAD, Packet Assembly/Disassembly Facility. A device that allows a dumb terminal to interface with a packet-switching network by performing the packet assembly and disassembly functions (ch. 9).

Pad Character. A series of alternating 0s and 1s usually sent at the beginning of a synchronous transmission to accomplish bit synchronization (ch. 3).

PAM. *See* **Pulse Amplitude Modulation.**

Parallel Interface. A hardware interface in which an entire character or group of bits is transmitted at one time by sending each bit in the character or group over a separate wire or pin (ch. 3).

Parity. An error-detection scheme based on the sum of all 1 bits, whereby one or more extra bits are added to help the receiving device detect an error (ch. 4). *See* **Vertical Redundancy Check, Longitudinal Redundancy Check,** and **Block Check Character.**

Parity, Even. One of the vertical redundancy check, error-detection schemes, whereby one extra bit is added to a group of bits such that the sum of all 1 bits is even (ch. 4).

Parity, Odd. One of the vertical redundancy check, error-detection schemes whereby one extra bit is added to a group of bits such that the sum of all 1 bits is odd (ch. 4).

Part 68. FCC rules that allow manufacturers to register their communications equipment with the FCC provided they meet FCC requirements designed to ensure the equipment does no harm to the telephone network. These rules are also referred to as the

FCC's registration program, and equipment that has been registered in accordance with these rules is called registered equipment and is assigned a registration number by the FCC (ch. 11).

Path. A particular route through a network consisting of one or more links (ch. 3).

PBX, Private Branch Exchange. A telephone-switching system that connects users' phones to each other, to central-office lines, or to special telephone lines, characterized by the user dialing specific numbers such as a "9" to get an "outside" or central office line (ch. 2).

PCM. *See* **Pulse Coded Modulation.**

PDM. *See* **Pulse Duration Modulation.**

Peripherals. Devices attached to a computer for input, output, or storage, such as terminals, printers, disk drives, or tape drives (ch. 1).

Permissive Device. A classification of registered modems whose output is limited to −9 dBm (ch. 5). *See* **Fixed Loss Loop Device.**

Personal Computer. A self-contained, desk-top or portable computer, whose user is responsible for its operations (ch. 1).

Phase Hits. A type of distortion that causes the phase of a signal to "jump" so there is a difference in phase between the transmitted and received signal (ch. 5).

Point to Point Network. *See* **Star Network.**

Poll. A message from the controlling station to a slave station asking if it has any message to transmit. If the slave terminals are connected to a device such as a cluster controller, the controlling station polls the cluster controller and either asks whether there are messages from any devices attached to the cluster controller (a general poll), or asks for messages from a specific device attached to the cluster controller (a specific poll) (ch. 6).

Poll, General. *See* **Poll.**

Poll, Specific. *See* **Poll.**

Polling. A method of policing traffic when multiple devices are connected to the same line, whereby the controlling station sends a message to the slave station asking if it has any message to transmit (ch. 6). *See* **Poll,** *See* **Selecting.**

Port Concentrator. A device that replaces a statistical or time division multiplexor at the computer end of the circuit but instead of breaking the data stream into separate data channels feeds the entire data stream into one computer port. The computer must be programmed to handle this multiplexed data stream from the port concentrator (ch. 8).

Port Selector. A device that allows a variety of other devices and types of lines to be connected to it and allows these devices to select which computer port they wish to be connected to (ch. 8).

Power Line Carrier. A technique that uses a radio frequency carrier transmitted over the AC power lines in a building to transmit data over those power lines (ch. 9).

Present Value. The value of various payments (or other economic transactions) made over a period of time translated into their value at the beginning of the time period and taking the time value of money (e.g. interest) into account (ch. 12).

Programmable Device. A classification of registered modems using a connection jack that can be programmed to set the output level based on the distance to the telephone company's central office (ch. 5). *See* **Fixed Loss Loop Device** and **RJ45S.**

Propagation Delay. The time it takes for a message to travel from the transmitting point to the receiving point on a circuit (ch. 6).

Protocol. The formal rules or conventions that govern the communications between two devices (ch. 4).

PSC. *See* **Public Utilities Commission.**

Public Utilities Commission (PUC); also known as a **Public Service Commission (PSC).** The state regulatory body that is the counterpart of the FCC for regulating communication services within a state (ch. 11).

PUC. *See* **Public Utilities Commission.**

Pulse Amplitude Modulation (PAM). An analog transmission technique that varies or modulates the height (amplitude) of each pulse in a series of pulses to correspond to the height of the waveform or data being transmitted (ch. 2).

Pulse Coded Modulation (PCM). A digital transmission technique whereby a waveform is sampled at equal time periods and the height of the waveform at the sample point (or the height of the pulse) is coded as a binary number, and that binary number (a series of 0s and 1s) is then transmitted (ch. 2).

Pulse Duration Modulation. *See* **Pulse Width Modulation.**

Pulse Width Modulation (PWM); Pulse Duration Modulation (PDM). An analog transmission technique that varies or modulates the width or duration of each pulse in a series of pulses to correspond to the height of the waveform or data being transmitted. *See* **Pulse Amplitude Modulation.**

PWM. *See* **Pulse Width Modulation.**

Radio Common Carriers. A specialized group of common carriers providing mobile radio service (ch. 11).

Regenerator. *See* **Repeator.**

Registered Equipment. *See* **Part 68.**

Registration. *See* **Part 68.**

Registration Number. *See* **Part 68.**

Repeator; Regenerator. An electronic device used in digital transmission that takes in attenuated, misshapen pulses and outputs them as perfectly shaped pulses of the correct height. The device is similar to an amplifier in an analog system but reshapes the pulse as well as correcting its height (ch. 2).

Ring. *See* **Loop.**

RJ11C/W. The USOC given to the standard type of modular jack used with a single line telephone (ch. 5).

RJ16X. *See* **RJ36X.**

RJ27X. *See* **RJ45S.**

RJ36X. The USOC given to the modular jack used with a single line telephone when employed for alternate voice data use. In this case the RJ16X modular jack is used for the modem (ch. 5).

RJ41S. The USOC code given to the modular jack known as a "universal data jack". It can accommodate modems classified either as programmable devices or fixed loss loop devices (ch. 5).

RJ45S. The USOC given to a special programmable data modular jack that can be programmed by inserting a resistor of the proper value. It is used with modems classified as "programmable devices." If multiple modems are being used a version known as the RJ27X is available that can connect up to eight modems (ch. 5).

Route. *See* **Path.**

RS-232C. The EIA standard for a twenty-five pin hardware interface that has become the most popular interface in the U.S. for use between a computer and a modem or computer terminals (ch. 3).

RS-366A. The EIA standard for the twenty-five pin interface between DTE and the Automatic Calling Unit (ACU) in auto-dial applications (ch. 3).

RS-422A. The EIA standard for voltage levels for the RS-449 interface when used at 20kpbs to 10mbps. This requires a balanced line (ch. 3).

RS-423A. The EIA standard for voltage levels for the RS-449 interface when used at 0 to 20kbps. This requires an unbalanced line (ch. 3).

RS-449. The EIA standard for an interface designed to replace the RS-232C. It provides for higher data rates and longer cable lengths than the RS-232C. It is used in conjunction with either the RS-422A or RS-423A standards (ch. 3).

RTS/CTS Delay. *See* **Training Period.**

SCC. *See* **Specialized Common Carriers.**

Select-fast. *See* **Selecting.**

Select-hold. *See* **Selecting.**

Select-verify. *See* **Selecting.**

Selecting. A technique used in conjunction with polling to tell a device that it is the device that will be used by the message to be transmitted. If the controlling station sends a select message and waits for the slave device to respond that it is ready to receive the message, that is called "select-hold" or "select-verify." If the controlling station sends a select message and then immediately sends the data, it is called "fast-select" (ch. 6). *See* **Polling.**

Selective Retransmission. *See* **ARQ, Continuous.**

Self-clocking. Applied when a device derives its "clock" or timing signal from the 0 and 1 transitions in the data stream (ch. 3).

Signal to Noise Ratio. A measurement of the amount of noise on the line relative to the signal power (ch. 5).

Simplex. A term used to describe a circuit, communication channel, protocol, or environment in which communication can only take place in one direction. However, the International Telecommunications Union defines "Simplex" as meaning that a signal can be transmitted in either direction but not simultaneously (ch. 4). *See* **Half Duplex** and **Full Duplex.**

Slotted Network Access. A method of policing traffic on a ring or loop whereby the transmission time is broken into equal length slots or time periods and each node can fill any empty slot that comes by (ch. 6).

SNA; Systems Network Architecture. An intelligent communications network approach developed by IBM (ch. 7).

SOH. A control character meaning "start of header" (ch. 4).

Space. A term used in asynchronous data transmission originating from its use in telegraphy; a start bit, a binary 0 (ch. 3). *See* **Mark.**

Specialized Common Carriers (SCC). *See* **Other Common Carrier.**

Star Network. Also known as a point to point network, this is characterized by a central control unit with a separate link to each device connected to the central control unit (ch. 6).

Stat MUX. *See* **Multiplexor, Statistical.**

Store and Forward System. An electronic message system that stores a message while or after it is created and later forwards it on to its destination. Store and forward systems can be used for electronic mail, written messages, voice messages, teletypewritten messages, FAX, etc (ch. 1).

SYN. A control character standing for synchronization. This character is used to get a data stream "in sync" (ch. 4).

Synchronous Data Link Control (SDLC). A subset of HDLC developed by IBM. SDLC is a bit-oriented, positional, synchronous transmission protocol intended to replace BISYNC in most applications (ch. 4).

Synchronous Transmission. A message oriented transmission scheme whereby the entire message is framed and each bit within the message is transmitted synchronously (or in sync) with a timing device called a "clock" (ch. 3).

Systems Network Architecture. *See* **SNA.**

Tariffs. The specific rules under which a common carrier offers service to the public. Tariffs amount to contracts between the customer and the common carrier (ch. 11).

TDM. *See* **Multiplexor, Time Division.**

Telenet. A public packet-switched network owned by GT&E (ch. 7).

Telex. An international teletypewriter service operating over the public telegraph networks at a speed of 50 bits/sec (ch. 1).

Token Passing. A method of policing traffic on ring and loop networks. A special message called a "token", consisting of a unique bit pattern, is placed on the ring and transmitted from node to node. When a node receives the token it can transmit any messages it has before retransmitting the token on to the next node. A form of polling whereby any station can send a poll to any other station is sometimes referred to as token passing on a bus (ch. 6).

Topology. The physical structure or arrangement of the nodes and links of the network (ch. 6).

Training Period; also **Training Time** or **RTS/CTS Delay.** The time between receiving the request to send the (RTS) signal and sending back the go-ahead to send the (CTS) signal that the modem with automatic equalizer uses to adjust its equalization parameters (ch. 5).

Training Time. *See* **Training Period.**

Tree. A multidrop or multipoint network topology that results from the connection of two or more star or bus networks at one of their nodes. Trees can be either "rooted" if there is a clearly defined bare node, or "unrooted" if there is not (ch. 6).

Tree, Rooted. *See* **Tree.**

Tree, Unrooted. *See* **Tree.**

Turnaround Time. With half-duplex transmission, the time it takes to change the directions of data transmission, which may include the time it takes to disable the echo suppressors (ch. 5).

20ma Current Loop. An interface widely used with teletypewriter terminals (ch. 3).

Twisted Pair. A pair of wires twisted at regular intervals to minimize certain types of noise when used for data communications (ch. 9).

TWX. A dial-up teletypewriter service operated by Western Union using analog transmission over the public telephone network at speeds up to 150 bits/sec (ch. 1).

USOC (Universal Service Ordering Code). A series of standard codes consisting of letters and numbers used by telephone companies to identify the various products they offer (ch. 5).

V.24. The CCITT equivalent of the RS-232C (ch. 3).

V.25. The CCITT equivalent of the RS-366A (ch. 3).

V.35. The CCITT recommended interface for higher speed applications than the V.24. The V.35 is used as part of CCITT's recommended X.25 packet-switching protocol (ch. 3).

Value-Added Carriers. Common carriers that used to lease lines from common carriers and added features to provide computer-based enhanced communications services such as packet switching (ch. 11).

Value-Added Networks. A type of digital network that not only transmits data but also offers code and speed conversion as well as other features (ch. 9).

Vertical Redundancy Check. An error-detection scheme whereby one extra bit (parity bit) is added to each group of bits such that the sum of 1s in the group plus the parity bit equals either an odd or even value depending upon which version is being used (ch. 4).

Video Display Terminal (VDT). A computer terminal with a keyboard for input and a CRT or Cathode Ray Tube (television-type tube) as an output or display device. Often called a CRT terminal or CRT for short (ch. 1).

Voice Communications. The process of transmitting information that originates as a human voice or as an imitation of the human voice and is intended to be received by another human or by computerized voice-recognition equipment (ch. 1).

Voice Grade Circuit. A telephone-company type circuit designed to transmit the human voice, in contrast to a circuit designed for telegraphy, data, or video transmission. Voice grade circuits are normally used with conventional modems to transmit data up to 9600 bps, or in some cases at 14,400 bps or at somewhat higher speeds (ch. 5).

X.21. The CCITT recommended interface for use between DTE and DCE operating on a public switched digital network (ch. 3).

X.21bis. The CCITT recommended interface for use on a public switched digital network as an interim solution, between DTE and DCE designed for the RS-232C or V.24 interfaces (ch. 3).

X.25. A CCITT recommended network access protocol for connecting DTE to public packet-switched networks (ch. 9).

X.75. A CCITT recommended protocol for the transference of data between two otherwise incompatible public packet-switched networks (ch. 9).

X-Out-of-Y Code. *See* **Constant Ratio Code.**

XON-XOFF. One of the major flow-control techniques that uses a special control character called XON to turn the transmitting device on, and one called XOFF to turn the transmitting device off (ch. 4).

INDEX

(Terms in italics also appear in the Glossary.)